Assessment of Deployment- and Mobilization-to-Dwell Policies for Active and Reserve Component Forces

An Examination of Current Policy Using Select U.S. Joint Force Elements

JOHN C. JACKSON, LISA M. HARRINGTON, JEFFREY S. BROWN, BRADLEY DEBLOIS, KATHERINE C. HASTINGS, JEANNETTE GAUDRY HAYNIE, DUNCAN LONG, MAX STEINER, JONATHAN WELCH, JOHN D. WINKLER

Prepared for the Office of the Secretary of Defense
Approved for public release; distribution unlimited

For more information on this publication, visit **www.rand.org/t/RRA670-1**.

About RAND
The RAND Corporation is a research organization that develops solutions to public policy challenges to help make communities throughout the world safer and more secure, healthier and more prosperous. RAND is nonprofit, nonpartisan, and committed to the public interest. To learn more about RAND, visit www.rand.org.

Research Integrity
Our mission to help improve policy and decisionmaking through research and analysis is enabled through our core values of quality and objectivity and our unwavering commitment to the highest level of integrity and ethical behavior. To help ensure our research and analysis are rigorous, objective, and nonpartisan, we subject our research publications to a robust and exacting quality-assurance process; avoid both the appearance and reality of financial and other conflicts of interest through staff training, project screening, and a policy of mandatory disclosure; and pursue transparency in our research engagements through our commitment to the open publication of our research findings and recommendations, disclosure of the source of funding of published research, and policies to ensure intellectual independence. For more information, visit www.rand.org/about/principles.

RAND's publications do not necessarily reflect the opinions of its research clients and sponsors.

Published by the RAND Corporation, Santa Monica, Calif.
© 2023 RAND Corporation
RAND® is a registered trademark.

Library of Congress Cataloging-in-Publication Data is available for this publication.

ISBN: 978-1-9774-0830-3

Cover: Bottom right: Vincent De Groot/U.S. Air National Guard Top left: Cpl. Karis Mattingly/U.S. Marine CorpsTop right: Cpl. Guy Mingo/U.S. Department of DefenseBottom left: Courtesy Photo/U.S. Marine Corps.

Limited Print and Electronic Distribution Rights
This document and trademark(s) contained herein are protected by law. This representation of RAND intellectual property is provided for noncommercial use only. Unauthorized posting of this publication online is prohibited. Permission is given to duplicate this document for personal use only, as long as it is unaltered and complete. Permission is required from RAND to reproduce, or reuse in another form, any of its research documents for commercial use. For information on reprint and linking permissions, please visit www.rand.org/pubs/permissions.

About This Report

Deployment-to-dwell (D2D) and mobilization-to-dwell (M2D) ratios were established during a period of extended operations in Iraq and Afghanistan. The security environment has subsequently changed to focus on great-power competition. After an extended period of operating under an "operational reserve" construct, involving protracted deployments and extended utilization of the reserve components, questions have been raised as to whether these policy goals remain relevant. The research described in this report had two primary objectives. The first was to examine implementation of existing D2D and M2D policies in each service under various scenarios. The second was to recommend changes and improvements to the policy to inform and optimize the U.S. Department of Defense's force utilization decisions.

The research reported here was completed in September 2021 and underwent security review with the sponsor and the Defense Office of Prepublication and Security Review before public release.

RAND National Security Research Division

This research was sponsored by the Office of the Secretary of Defense and conducted within the Personnel, Readiness, and Health Program of the RAND National Security Research Division (NSRD), which operates the National Defense Research Institute (NDRI), a federally funded research and development center sponsored by the Office of the Secretary of Defense, the Joint Staff, the Unified Combatant Commands, the Navy, the Marine Corps, the defense agencies, and the defense intelligence enterprise.

For more information on the RAND Personnel, Readiness, and Health Program, see www.rand.org/nsrd/prh or contact the director (contact information is provided on the webpage).

Acknowledgments

We would like to thank Judd Lyons, Office of the Under Secretary of Defense for Personnel and Readiness, Reserve Integration, for providing direction for the joint force element modeling supporting this analysis. We would also like to thank the action officers for the study, COL Chad Bridges and Col James Gresis, for their oversight and insights. We thank our RAND Corporation colleague Barbara Bicksler for her assistance in improving the clarity of the manuscript and the quality of presentations delivered to the sponsor throughout the study. Finally, we greatly appreciate the insightful feedback of our reviewers, Michael Linick and David Chu.

Summary

Deployment-to-dwell (D2D) and mobilization-to-dwell (M2D) ratios were established during a period of extended operations in Iraq and Afghanistan. In a November 2013 memorandum revising these guidelines, the Under Secretary of Defense for Personnel and Readiness stated,

> The intent is for commanders at every level to ensure individual service members, regardless of unit assignment, are not repeatedly exposed to combat, do not experience disproportionate deployments, and do not spend extended periods of time away from their homeport/station/base unless required by operational necessity.[1]

After a lengthy period of operating under an "operational reserve" construct, involving protracted deployments and increased utilization of the reserve components, questions have been raised as to whether these policy goals remain relevant—particularly because the security environment has evolved to focus on great-power competition.

The research described in this report had two primary objectives. The first was to examine implementation of existing active component D2D and reserve component M2D policies in each service under various scenarios. The purpose of this analysis was to better understand the relationship between the D2D/M2D policy and readiness, with an emphasis on the operational impacts of the policy. The second objective was to recommend changes and improvements to the policy to inform and optimize the Department of Defense's (DoD's) force utilization decisions.

Research Approach

A fundamental understanding of how each of the services approaches fulfilling the D2D/M2D policy was essential prior to examination of the policy under possible future scenarios. To this end, we conducted extensive meetings with each service's force management organization to understand service-specific policies supporting D2D and M2D (if any were in place), how each service manages D2D and M2D, and the extent to which the services use modeling to inform their management. We also interviewed force management entities on the Joint Staff to ensure that this work captured approaches to D2D/M2D policy implementation specific to the joint force.

After establishing a foundational understanding of how the services approached supporting D2D/M2D policy, we adapted existing analytic models to evaluate the ability of the ser-

[1] Jessica L. Wright, Acting Under Secretary of Defense for Personnel and Readiness, "Under Secretary of Defense (Personnel & Readiness) Deployment-to-Dwell, Mobilization-to-Dwell Policy Revision," memorandum for Secretaries of the Military Departments and Chairman of the Joint Chiefs of Staff, Washington, D.C., November 1, 2013.

vices to meet operational demands with active and reserve component forces using four illustrative joint force elements (JFEs). Each of the JFEs, which were selected in conjunction with the research sponsor for analysis, is a unit or platform that frequently deploys and plays a role in responding to major contingency operations. The four JFEs were the U.S. Army's armored brigade combat team, the U.S. Air Force's (USAF's) KC-135 fleet, the U.S. Marine Corps' infantry battalion, and the U.S. Navy's aircraft carrier. The analysis of each JFE relied on the latest information available from each service on its approach to readiness (including specific readiness models where applicable) and is grounded in recent employment demand for each of the force elements. Modeling each force element under various assumptions provided insight on the capacity of the force element to fulfill demands, illuminated consequences of a given set of demands, and led to recommendations for improvements to the policy.

Observations

Major observations from this study that led to the recommendations are as follows:

- The operational context for D2D/M2D policy has changed dramatically since the inception of the original D2D/M2D policy; the challenges currently facing DoD are fundamentally different from those faced in Iraq and Afghanistan.
- All Secretary of Defense–approved operations count in calculating D2D/M2D ratios regardless of service member exposure to combat conditions.
- DoD views all Secretary of Defense–approved operations as consuming readiness, but some operations actually improve readiness.
- The services have differing views as to what counts as a deployment.
- Variations in service requirements lead to different methods for managing force utilization that are not confined to unit and individual D2D/M2D ratio calculations.
- Administrators of the D2D/M2D policy in the Office of the Secretary of Defense state that, in the future, an intended outcome of the policy is to help improve operational readiness for a return to great-power competition.

Recommendations

Our investigation concluded with three recommendations: (1) clarify the purpose of the D2D/M2D policy, (2) consider redefining what counts as a deployment, and (3) review how D2D and M2D are managed.

Clarify the Purpose of the Policy Within the Current Operational Context

The intent of the original D2D/M2D policy was to prevent the overexposure of individuals to combat, yet the operational environment has changed since the policy was developed; fewer service members are exposed to combat conditions as frequently. Even so, the services track all operational deployments as part of their D2D/M2D calculations rather than focusing on only those that result in combat exposure. In addition, departmental discussions often include preserving unit readiness as a goal of the policy. The written goals of the D2D/M2D policy and how it is being used may be diverging over time. Therefore, we recommend that DoD revisit the purpose of the policy so that it fully suits DoD needs.

Consider Redefining What Counts as a Deployment

Depending on how the services apply the current policy, the D2D/M2D ratios of some units might be artificially inflated and, for other units, might lead to false assumptions on available capacity for responding to a major contingency operation. Moreover, not all deployments consume readiness; some deployments enhance readiness. With a return to great-power competition and renewed focus on readiness required to respond to a major contingency, D2D/M2D policy could allow the services to make judgments about excluding readiness-enhancing deployments from D2D/M2D calculations. As of the writing of this report, DoD is considering changes in what counts as a deployment.

Review How Deployment to Dwell and Mobilization to Dwell Are Managed Across the Department of Defense

This research revealed that the services have different ways of managing the force that are based in part on D2D/M2D policy; the Army and Marine Corps generally focus on the unit, while the USAF and Navy manage the individual and the platform. Because of these differences, to fully understand the impact of deployments on operational readiness, D2D/M2D ratio calculation, tracking, and corresponding reporting requirements should consider platform, crew, and unit composition.

Finally, if the purpose of D2D/M2D policy remains as written—to avoid the overexposure of individuals to operational deployment—then the recently revised DoD personnel tempo policy could suffice in fulfilling this purpose. However, if the purpose of the policy is to preserve operational readiness (i.e., readiness required to support a major contingency operation), DoD should consider D2D/M2D tracking only for those JFEs and other essential force elements needed for response to major contingency operations. Such a move would dramatically decrease the complexity of D2D/M2D tracking and provide greater clarity into the effects that emerging requirements have on DoD's ability to respond to major contingencies.

Contents

Figures and Tables

Figures

Tables

CHAPTER ONE

Introduction

When the U.S. Department of Defense (DoD) established utilization guidelines for active and reserve component forces, it did so in response to congressional pressure to ensure that service members had sufficient time at home. In the late 1990s, the shrinking post–Cold War force structure was primarily used in support of peacekeeping and humanitarian operations, and overuse was considered to have negative effects on readiness, family, and retention. These concerns persisted but evolved in the aftermath of the attacks on the Pentagon and the World Trade Center on September 11, 2001. Ultimately, in 2007, DoD established deployment-to-dwell (D2D) and mobilization-to-dwell (M2D) ratios that govern active component deployments and reserve component mobilizations and assist with sizing and distributing capabilities within components.

The goal of these utilization guidelines was to reduce strain on the force caused by excessive deployments. In a November 2013 memorandum revising these guidelines, the Under Secretary of Defense for Personnel and Readiness (USD P&R) stated goals for the guidelines:

> The intent is for commanders at every level to ensure individual service members, regardless of unit assignment, are not repeatedly exposed to combat, do not experience disproportionate deployments, and do not spend extended periods of time away from their homeport/station/base unless required by operational necessity.[1]

According to policy goals, a one-year period of deployment for active component forces should be followed by a two-year period of dwell (1:2), and a one-year period of mobilization for reserve component personnel should be followed by a five-year period of demobilization (1:5).

However, even at the time that the memo was written, the Secretary of Defense acknowledged that force needs in Iraq and Afghanistan would lead to a near-term operational tempo (OPTEMPO) ratio of 1:1 for the active component and 1:4 for the reserve component, referred to as *threshold ratios.*

---

[1] Jessica L. Wright, Acting Under Secretary of Defense for Personnel and Readiness, "Under Secretary of Defense (Personnel & Readiness) Deployment-to-Dwell, Mobilization-to-Dwell Policy Revision," memorandum for Secretaries of the Military Departments and Chairman of the Joint Chiefs of Staff, Washington, D.C., November 1, 2013.

Operations in Iraq and Afghanistan were conducted under an "operational reserve" construct, involving protracted deployments and extended utilization of the reserve components. However, the security environment has since changed to focus on great-power competition, and questions have been raised as to whether these dwell ratio planning guidelines remain relevant. In this report, we begin to explore these questions by examining how utilization guidelines are currently implemented in each of the military services, paying particular attention to how the policy can be improved in the long run.

Comparison of the Deployment-to-Dwell/Mobilization-to-Dwell Policy: Past and Present

The modern history of the D2D ratio begins in the late 1990s with the passage of 10 U.S.C. 991.[2] This section of law, which was passed by Congress in 1999 and took effect in 2000, was a response to congressional concern over increasing demands on some U.S. military members. The rise in peacekeeping and humanitarian missions at a time of declining military manpower in the first decade of the post–Cold War era had led to some units deploying at rates far higher than the Cold War norm. Congress worried that the pressure of a high-deployment tempo would increase stress on military families while affecting readiness and retention.[3]

The effect on readiness was complicated. Looking at data from U.S. Army units deployed to Bosnia, U.S. European Command postulated a nonlinear, bell-shaped impact of OPTEMPO on individual and unit readiness.[4] An ideal level of limited deployments generated "peak" readiness, while OPTEMPO that was too high or too low had a negative effect on readiness.[5] With no deployments, units were insufficiently practiced on their mission-essential task lists. With too much time spent on deployments, units had insufficient training time to address mission-essential task requirements not used during their deployments. There was no such bell-curve effect on medical readiness; increasingly longer deployments adversely affected psychological and physical readiness.[6]

---

[2] U.S. Code, Title 10, Armed Forces; Subtitle A, General Military Law; Part II, Personnel, Chapter 50, Miscellaneous Command Responsibilities; Section 991, Management of Deployments of Members and Measurement and Data Collection of Unit Operating and Personnel Tempo, January 1, 2021.

[3] Lawrence Kapp, *Recruiting and Retention in the Active Component Military: Are There Problems?* Washington, D.C.: Congressional Research Service, RL31297, February 25, 2002.

[4] OPTEMPO is the pace at which units are involved in military activities, including deployments, exercises, and training activities.

[5] Carl A. Castro and Amy B. Adler, *The Impact of Operations Tempo: Issues in Measurement*, Ft. Detrick, Md.: U.S. Army Medical Research and Material Command, 2000.

[6] Carl A. Castro and Amy B. Adler, "OPTEMPO: Effects on Soldier and Unit Readiness," *Parameters*, Vol. 29, No. 3, Autumn 1999.

In terms of family stress, deployment rates were seen as problematic. A late-1990s Army study of forces in U.S. European Command found that, among the 1,305 soldiers with families surveyed, 61.7 percent of those who intended to leave the military after fulfilling their service obligations said that the number of deployments caused a big strain on the family. Even among those who intended to stay past their current service obligations, 54 percent said the deployment pace caused family strain, an opinion shared by 49 percent of those intending to stay at least until retirement.[7]

The impact of deployments on retention was mixed. One-third of soldiers who intended to leave the military reported that they were doing so because of the deployment tempo.[8] However, a contemporaneous RAND Corporation study found that a limited number of short (three-month) deployments increased retention—although longer deployments had a negative impact.[9]

Together, these findings prompted a congressional response. Section 991 had three main objectives: (1) to establish one- and two-year high-deployment thresholds, (2) to instruct the Secretary of Defense to create and transmit to Congress a D2D policy for the active component and an M2D policy for the National Guard and reserve, and (3) to develop a system to measure unit OPTEMPO and personnel tempo (PERSTEMPO)—the amount of time military members spend away from their permanent housing—for all personnel.

Shortly after Section 991 took effect, the 9/11 attacks led DoD to invoke the national security exemption on deployment thresholds written into the legislation. This waived the requirement to keep deployment ratios within the limits that Congress had established two years prior. To compensate deployed military members for longer and more frequent deployments, DoD also implemented allowances for Imminent Danger Pay and Hardship Duty Pay as an alternative to the high-deployment per diem or high-deployment allowance under 37 U.S.C. 436. These location-based allowances were both easier to administer and less costly than the more granular (and generous) $100 per diem originally proposed by Congress.[10]

During this time, DoD reported PERSTEMPO but focused primarily on measuring and optimizing OPTEMPO. In the 2001 Quadrennial Defense Review, DoD had identified a need to develop "realistic tempo standards and limitations to control explicitly the amount of time DoD personnel are deployed away from home station or stationed outside the United States."[11] Many of the recommendations of this review, published on September 30, 2001, were deprioritized immediately with the concurrent beginning of the global war on terror.

---

[7] Kapp, 2002.

[8] Castro and Adler, 1999.

[9] James Hosek, *Perstempo: Does It Help or Hinder Reenlistment?* Santa Monica, Calif.: RAND Corporation, RB-7532, 2004.

[10] U.S. Code, Title 37, Pay and Allowances of the Uniformed Services; Chapter 7, Allowances; Section 436, High-Deployment Allowance: Lengthy or Numerous Deployments; Frequent Mobilizations, 1999.

[11] DoD, *Quadrennial Defense Review Report*, Washington, D.C., September 30, 2001, p. 59.

The beginning of large-scale combat operations in Iraq in 2003 introduced further stresses on the deployment schedule of military personnel. The first major administrative change was a shift in what counted as a deployment. The metric defined in 10 U.S.C. 991—"any day on which, pursuant to orders, the member is performing service in a training exercise or operation at a location or under circumstances that make it impossible or infeasible for the member to spend off-duty time in the housing in which the member resides"—was put aside.

In place of this metric, the military adopted the concept of *boots on ground* (BOG). BOG referred to the time that soldiers spent in theater at their deployed locations. BOG time started when the majority of a unit (defined as the unit's *center of mass*) arrived in theater and ended when at least half the unit had left.[12] The adoption of BOG as a force management construct had two major effects. First, it narrowed the definition of what counted as *deployed* from unable "to spend off-duty time in the housing in which the member resides"[13] to "time in theater."[14] Second, it shifted attention from the individuals to the unit as the focus for measuring the frequency and duration of deployments. However, the appropriate unit level was left to the services, with the Army measuring BOG for brigades and regiments and the Marine Corps measuring BOG for battalions and squadrons.[15]

In 2003, the Secretary of Defense issued a memorandum on the utilization of reserve component forces. This memo set a planning objective that limited involuntary mobilizations for these forces to a ratio of 1:5.[16]

In 2007, the Secretary of Defense issued a memorandum on the utilization of the total force, which established departmental planning objectives for deployment ratios across the services.[17] The memo formalized a D2D goal of 1:2 for the active component and 1:5 for the reserve component.[18] The memo also stated that deployments would be measured at the unit level, which codified the then-current practice first described under the BOG policy but differed from the focus on individuals found in 10 U.S.C. 991 as originally drafted.

In 2011, Congress updated the definition of *dwell time* in Section 522 of the fiscal year (FY) 2012 National Defense Authorization Act, defining it as time spent "at the perma-

---

[12] Timothy M. Bonds, Dave Baiocchi, and Laurie L. McDonald, *Army Deployments to OIF and OEF*, Santa Monica, Calif.: RAND Corporation, DB-587-A, 2010.

[13] 10 U.S.C. 991.

[14] David Chu, Under Secretary of Defense for Personnel and Readiness, "Measuring Boots on Ground (BOG)—Snowflake," memorandum for the Secretary of Defense, Washington, D.C., November 22, 2004.

[15] Chu, 2004.

[16] John D. Winkler, "Developing an Operational Reserve: A Policy and Historical Context and the Way Forward," *Joint Forces Quarterly*, Vol. 59, 4th Quarter 2010, p. 16.

[17] Robert M. Gates, Secretary of Defense, "Utilization of the Total Force," memorandum for the Secretaries of the Military Departments, Chairman of the Joint Chiefs of Staff, and Under Secretaries of Defense, January 19, 2007.

[18] Gates, 2007.

nent duty station or home port after returning from a deployment."[19] OPTEMPO and PERSTEMPO were also redefined, and record-keeping requirements were updated. Additionally, the FY 2012 National Defense Authorization Act allowed the Secretary of Defense to modify the definition of *dwell* if Congress was kept informed.

In 2013, USD P&R published a memorandum on D2D and M2D policies that made a minor adjustment to the OPTEMPO definition and established goal and threshold dwell ratios.[20] Operational deployments were defined as starting when the majority of a unit left its home station or en route training location in support of Secretary of Defense–approved operational plans, concept plans, or Executive orders.[21] D2D ratios were measured at the unit level, and units were either "on deployment" or "in dwell." The goal ratios established the ideal minimum D2D/M2D ratios used to guide the force generation process and projected requirements. The *threshold ratio* is the deployment rate above which Secretary of Defense approval is required to deploy. The active component would target a 1:2 D2D ratio (twice as much dwell time as deployment time), and no active component unit could exceed a 1:1 ratio. The reserve and Guard components would target a 1:5 M2D ratio with a 1:4 threshold ratio.

In 2021, USD P&R issued a memorandum updating the 2013 policy guidance.[22] The 2021 guidance updated the threshold and planning objective ratios for the active component force. While the reserve and Guard M2D dwell ratios remain 1:5 (goal) and 1:4 (threshold), the active duty ratios have been lengthened to 1:3 (goal) and 1:2 (threshold). This update reflects the reduction in force requirements with the ongoing drawdown from global war on terror deployments. However, DoD retains the statutory authority—given to it by Congress through 10 U.S.C. 991—to further adjust D2D/M2D ratios if national security considerations change.

Table 1.1 summarizes the changing definitions of *deployment* and *dwell*, the levels at which deployments and dwell time have been tracked, and the goal and threshold ratios for military force generation processes between the passage of 10 U.S.C. 991 and the 2021 guidance.

Research Objectives

This research had two primary objectives. The first was to examine implementation of existing active component D2D and reserve component M2D policies in each service under various scenarios. More specifically, we examined the relationship between the D2D/M2D policy

---

[19] Public Law 112–81, National Defense Authorization Act for Fiscal Year 2012, December 31, 2011. See also Wright, 2013.

[20] Wright, 2013.

[21] Wright, 2013.

[22] Office of the Under Secretary of Defense, "Under Secretary of Defense (Personnel & Readiness) Deployment-to-Dwell, Mobilization-to-Dwell Policy Revision," memorandum for Secretaries of the Military Departments and Chairman of the Joint Chiefs of Staff, Washington, D.C., November 1, 2013.

TABLE 1.1
Comparison of Deployment-to-Dwell/Mobilization-to-Dwell Policy and Statute

	10 U.S.C. 991	"Utilization of the Total Force" Memo (Secretary of Defense, 2007)[a]	D2D/M2D Policy (USD P&R 2013)[c]	D2D/M2D Policy (USD P&R 2021)[d]
Deployment	A *deployment* is defined as "any day on which, pursuant to orders, the member is performing service in a training exercise or operation at a location or under circumstances that make it impossible or infeasible for the member to spend off-duty time in the housing in which the member resides."	"At Service discretion, [the deployment/ mobilization period] may exclude individual skills training required for deployment and post-mobilization leave."	"An operational deployment begins when the majority of a unit or detachment, or an individual not attached to a unit or detachment, departs . . . to meet a Secretary of Defense-approved operational requirement."	A deployment begins when a unit, detachment, or individual not attached to a unit or detachment leaves a homeport, station, or base or leaves from an en route training location to meet a Secretary of Defense–approved operation to meet an operational requirement.
Unit of measure for D2D/M2D calculation	Individual	Unit	Unit	Unit
Dwell	"The term 'dwell time' means the time a [regular] member of the armed forces or a unit spends at the permanent duty station or home port after returning from a deployment" or the amount of time "a reserve member of the armed forces remains at the member's permanent duty station after completing a deployment of 30 days or more in length."	Definition not addressed in memorandum	"Dwell is defined as the period of time a unit or individual is not on an operational deployment. . . . For the Reserve Component, dwell is defined as the period of time an individual is not mobilized" (includes pre- and postdeployment activities).	*Dwell* is the period of time the majority of a unit is not on an operational deployment. An active component unit is either on deployment or on dwell.

Table 1.1—Continued

	10 U.S.C. 991	"Utilization of the Total Force" Memo (Secretary of Defense, 2007)[a]	D2D/M2D Policy (USD P&R 2013)[c]	D2D/M2D Policy (USD P&R 2021)[d]
Thresholds	• One-year: No more than 220 days deployed out of the previous 365 days (1:0.66) • Two-year: No more than 400 days deployed out of the previous 730 days (1:0.825)	• Active component: No threshold, but a goal of 1:2 and a recognition that the 2007 status quo was 1:1[b] • Reserve component: Deployment length capped at 12 months. Goal remained 1:5, and there was a recognition that many units were being deployed more frequently in 2007	• Active component: 1:1 with a goal of 1:2 • Reserve component: 1:4 with a goal of 1:5	• Active component: 1:2 with a goal of 1:3 • Reserve component: 1:4 with a goal of 1:5

NOTES: SECDEF = Secretary of Defense.
[a] Gates, 2007.
[b] Reviewer comments for this report indicated that, during this time, considerable controversy was created by an Army effort to give active forces generous dwell times relative to reserve component units by increasing the goal ratio for the active component to 1:3. The reviewer went on to share that the Office of the Secretary of Defense expressed the view that this effort showed a willingness to sacrifice the reserve component to protect the active component.
[c] Wright, 2013.

and operational readiness. The second objective was to recommend changes and improvements to the policy to inform and optimize DoD's force utilization decisions.

Research Approach

We conducted this analysis using a case study approach for each service, at the request of the sponsor. A fundamental understanding of how each of the services approaches fulfilling the D2D/M2D policy was an essential starting point for our analyses and provided the foundation for each of the service case studies. To this end, we conducted extensive meetings with each service's force management organizations to understand whether service-specific D2D/M2D policies existed, how each service manages D2D and M2D, and to what extent the services use modeling to inform force management. We also had meetings with force management entities on the Joint Staff to incorporate joint force implementation approaches to D2D/M2D policy.

Once we had developed a foundational understanding of how the services approached the D2D/M2D policy, joint force elements (JFEs) were selected, in conjunction with the sponsor, for each service to evaluate how this policy affected the ability of the active and reserve components to meet operational demands under various scenarios. Each of the JFEs selected is a unit or platform that frequently deploys and plays a role in responding to major contingency operations. JFE representation in the reserve component was an important criterion for inclusion in this work. The JFEs selected were the armored brigade combat team (ABCT) for the U.S. Army, the infantry battalion for the U.S. Marine Corps, the KC-135 refueling fleet squadron for the U.S. Air Force (USAF), and the aircraft carrier for the U.S. Navy.

The models used to investigate the operational impacts of the D2D/M2D policy leverage the latest information available from each service on its approach to readiness (including specified readiness models where applicable) and are grounded in recent employment demand for each of the JFEs. These models are restricted to evaluating JFE availability during the competition period. The accompanying analysis for each of the JFEs provides insight into how each service manages D2D and M2D and associated implications of the policy for the operational readiness (i.e., readiness required to support a major contingency operation) of the active and reserve components during competition. These results assisted in identifying consequences of a given set of demands and possible recommendations for improvements to the ratios stated in the policy.

Organization of This Report

Chapters Two through Five contain the results of our investigation of D2D/M2D policies for case studies of the U.S. Army, U.S. Marine Corps, USAF, and U.S. Navy, respectively. The chapters begin with overviews of the service policies and how they are managed before turning to the results of the JFE analyses, which provide insight into how constraints imposed by the D2D/M2D policy affect the ability of the services to meet demands for forces. In Chapter Six, we summarize six dimensions of D2D/M2D policy drawn from the service analyses and propose recommendations for the policy.

CHAPTER TWO

Deployment- and Mobilization-to-Dwell Policy in the U.S. Army

In this chapter, we examine the Army's approach to implementation of the D2D/M2D policy using a high-interest JFE, the ABCT, as a case study. The analysis provides insight into how the Army manages D2D and M2D for its units and implications of the policy on the readiness of the active and reserve components during competition. Before describing the model, assumptions, and results of our ABCT analysis, the chapter begins with an overview of how the D2D/M2D policy is implemented in the Army. Special attention is paid to its role in future force structure decisions and the relationship between the policy and readiness.

Policy

The Army does not have a D2D/M2D policy that is separate from policy guidance provided by the Office of the Secretary of Defense (OSD). The Army's threshold (or "redline") ratio for active component forces has been 1:2, and the ratio for reserve component forces has been 1:4.[1] To deploy a unit that does not meet the threshold for dwell or demobilization requires a waiver signed by the Secretary of Defense. The goal ratio for active component forces is 1:3, and the objective for reserve component forces is 1:5.

The Army tracks D2D and M2D at the unit level. A *unit* is a "reporting unit" as described in Army Regulation (AR) 220-1, the Army regulation on readiness reporting.[2] It includes operating force units that are denoted with unit identification codes ending in -AA or -FF.[3] Units with -AA codes can range in size from teams of fewer than ten personnel to battalions with closer to 1,000 personnel. Units with -FF codes, or *composite reporting units*, contain one or more -AA units. The Army's brigade combat teams are -FF units.

---

1 This is according to Secretary of Defense contingency planning guidance.

2 AR 220-1, *Army Unit Status Reporting and Force Registration—Consolidated Policies*, Washington, D.C.: Department of the Army, April 15, 2010, p. 38.

3 Department of the Army Pamphlet 220-1, *Defense Readiness Reporting System—Army Procedures*, Washington, D.C., Headquarters, Department of the Army, November 16, 2011.

As per OSD policy, an active component unit is considered deployed when the majority of the unit is engaged in a Secretary of Defense–approved operational requirement away from home station. A reserve component unit is considered mobilized per the terms of Secretary of Defense–approved orders.

The Army tracks unit D2D/M2D status using the Army Synchronization Tool (AST).[4] AST is "the authoritative system of record to synchronize the Army's force generation process."[5] Staff at U.S. Army Forces Command (FORSCOM) indicate that AST is typically accurate and sufficient for planning, but they also occasionally verify reported D2D/M2D status captured in AST by checking the Mobilization and Deployment Information System for records of past deployments.[6] When making near-term sourcing decisions, they consult unit commanders both on reported unit readiness and on dwell.

Deployment-to-Dwell/Mobilization-to-Dwell Management

D2D/M2D status is a critical component of assessing unit availability for fulfilling emerging requirements and in managing known requirements. Known requirements are handled through the Global Force Management Allocation Plan (GFMAP). The GFMAP planning process is the deliberate means by which DoD adjudicates how forces are sourced to meet future requirements for forces. Such requirements are mainly the demands of combatant commands and their constituent service component commands—forces for ongoing named operations, presence, security cooperation activities, and the like—but also include forces held ready for potential contingency response. After combatant command requirements are validated, they are assigned to joint force providers to be sourced with specific units. For the Army, FORSCOM is principally responsible for this sourcing. In doing so, FORSCOM considers the current and projected D2D/M2D status of units to determine their availability to meet given requirements. The D2D/M2D status of units chosen to source a requirement is logged in Annex A of the GFMAP, a continuously updated record of requirements and sourcing decisions.

D2D and M2D are also used by FORSCOM to manage unforeseen requirements not captured in the GFMAP process.[7] When confronted with such a demand, FORSCOM planners first consider whether they have the authority to use the reserve component. Their second consideration is the D2D/M2D status of those units that can provide the required capability. If no unit has sufficient dwell and a waiver is needed, candidate units' readiness status is

[4] This is formerly known as the *Army Force Generation Synchronization Toolset*.

[5] Army Regulation 525-29, *Force Generation—Sustainable Readiness*, Washington, D.C.: Headquarters, Department of the Army, October 1, 2019, p. 8.

[6] Telephone discussion, U.S. Army Forces Command, August 17, 2020.

[7] Telephone discussion, U.S. Army Forces Command, November 14, 2019.

decisive in fulfilling the requirement; the unit that possesses the highest level of readiness is selected for deployment.

FORSCOM planners report that D2D and M2D seldom, if ever, prevent the Army from sourcing a requirement.[8] They also report requesting very few waivers to permit units to deploy below the D2D/M2D threshold.

The Role of Deployment-to-Dwell/Mobilization-to-Dwell Policy in Future Army Force Structure Decisions

The Army uses D2D/M2D policy to inform its force structure decisions. The Army plans its force structure—the numbers and types of units that it will man, train, and equip—through the Total Army Analysis (TAA) process. In broad strokes, every year through TAA, the Army compares the current plan for force structure with forecast demand. Where shortfalls are identified, senior leaders make decisions about whether and how to change structure within budget, end strength, and other constraints.[9]

In the most recent TAA (TAA 2023–2027, which refers to the program objective memoranda years through which force changes are planned), the Army used D2D/M2D thresholds to measure the share of its force that was available to meet steady-state demand.[10] Steady-state demand comprises planned deployments. For its analysis, the Army derived future steady-state demand from recent historical demand captured in the GFMAP.[11] If that future demand for a given unit type exceeded one-third of the active component inventory (reflecting a 1:2 threshold) and one-fifth of the reserve component inventory (reflecting a 1:4 threshold), then a shortfall was registered, and the unit type became a candidate for growth in structure. For this TAA, only a handful of the Army's 300-plus unit types failed to meet demand within the D2D/M2D threshold.

Although D2D/M2D thresholds were an important force management input to the Army's central force structure planning analysis, two features must be noted. First, as acknowledged by Army staff, use of the GFMAP as a proxy for future steady-state demand is imperfect. Barring sharp changes in the global operating environment, it is reasonable to assume that the near future will present the Army with similar demands as the recent past. However, that GFMAP demand—the demand actually sourced by Army forces in the GFMAP—was itself a product of the very D2D/M2D thresholds being used in the Army's analysis. Therefore, it is unsurprising that almost all Army force elements met steady-state demand in TAA, because

[8] Telephone discussion, U.S. Army Forces Command, November 14, 2019.

[9] Telephone discussion, G-3 Force Management, August 4, 2020.

[10] Telephone discussion, G-3 Force Management, August 4, 2020.

[11] Telephone discussion, G-3 Force Management, August 4, 2020.

the source of that demand took into account the same thresholds used to judge steady-state sufficiency.

Second, sufficiency to meet steady-state demand in TAA is significantly less important in Army decisions to resource changes in structure than is sufficiency to meet surge, or contingency, demand. After assessing steady-state sufficiency, TAA planners analyze the ability of the Army to meet demand for one or more large, unexpected conflicts against a peer competitor. This analysis does not consider D2D/M2D thresholds. Dwell is not considered a constraint for meeting contingency demand, and the great majority of the Army's inventory of forces is made available to fight. Assessed shortcomings in the conflict requirements analysis have much greater weight than do shortcomings in the steady-state analysis.[12] Of course, it is possible that a unit type will be found lacking in both assessments.

The Relationship Between Readiness and Deployment to Dwell/Mobilization to Dwell in the Army

As alluded to in the previous section, D2D/M2D policy is only one factor in Army force management. Another critical input is unit readiness. The two are related but independent. That is, current D2D/M2D policy does not drive unit readiness. The Army's approach to generating ready forces is referred to as *sustainable readiness*.[13] Sustainable readiness includes a concept called *unit readiness cycles* (URCs), which define for a period of time the intended progressive readiness status of a given unit.[14] While the system inherently allows for tailored URCs, the typical 24-month URC for an active component unit targets 12 months at C1, the highest readiness level.[15] The typical cycle for a reserve component unit creates a year of mission readiness out of a five-year cycle.[16]

Unit readiness generated according to these URCs (which, it should be emphasized, are not guarantees of actual readiness-generation performance) is not impeded by the current D2D/M2D thresholds of 1:2 and 1:4. The reserve component URC matches the M2D ratio of 1:4; when a unit meets its dwell target, it should also be on schedule to achieve mission readiness. An active component unit that achieves 12 months of C1 in a 24-month period is ready to deploy on a cycle compatible with a 1:1 D2D ratio and can easily meet readiness timelines under a 1:2 deployment schedule. At 1:2, D2D constrains availability of units, not readiness.

---

[12] Telephone discussion, G-3 Force Management, August 4, 2020.

[13] Army Regulation 525–29, 2019; and Headquarters, Department of the Army, *Regionally Aligned Readiness and Modernization Model (ReARMM)*, Washington, D.C., forthcoming (a force generation and readiness publication that is currently being drafted by the Army).

[14] Army Regulation 525–29, 2019, p. 24.

[15] Army Regulation 525–29, 2019, p. 28.

[16] Although mission ready, the reserve component unit might not achieve C1 or might not sustain it for the full year.

We hasten to note, however, that this distinction between readiness and unit availability does not indicate dysfunction or inefficiency. The Army is charged with having units ready for unexpected contingencies, not just steady-state commitments. Forces that are mission ready but D2D-constrained do not necessarily indicate "wasted" readiness.

Modeling Armored Brigade Combat Team Availability

Although simple arithmetic can describe the on-average ability of Army force elements to meet demand or required forces needed to meet peak demand for a given scenario, it disguises the inevitable variation in unit availability over time. The RAND Arroyo Center has developed a model that captures how D2D/M2D policy, readiness models, and other force management policies or alternative demands affect unit availability over time. By specifying policies and particular Army decision rules, it is possible to evaluate how demand is sourced over time, whether demand is met, and how well suited undeployed units are for fulfilling future demand. In the following sections, we describe the Multi-Period Assessment of Force Flow (MPAFF) tool and apply it to one potential future competition scenario for the ABCT.[17]

How the Model Works

The logic supporting the MPAFF tool is shown in Figure 2.1. The user can specify steady-state or contingency demands, or both, that manifest in the number of units required monthly over time and has the freedom to vary these requirements for any demand signal that they want to explore. In concert with the specified demand profile, the user defines the supply of forces and the policies that define the rules for the allocation of ready forces. Once these supply and demand constraints are in place, deployments are assigned to available units as they move through their readiness cycles. Correspondingly, previously deployed units proceed through a reset period, according to policy, during which they are unavailable to fill demand. MPAFF combines this information to track unit availability and demand that is met and unmet temporally, ultimately providing a way for decisionmakers to assess the operational effects of policy decisions, such as D2D and M2D. The tool captures the variability that policy and supply constraints have on unit availability in a way that such metrics as average D2D and M2D or static "peak-demand" scenarios do not.

Model Inputs

Modeling unit availability for any force element in the Army is enormously complex. Variables in the RAND model capture all dimensions of force generation for the Army. The foun-

---

[17] Katharina Ley Best, Igor Mikolic-Torreira, Rebecca Balebako, Michael Johnson, Trung Tran, and Krista Romita Grocholski, *Assessing Force Sufficiency and Risk Using RAND's Multi-Period Assessment of Force Flow (MPAFF) Tool*, Santa Monica, Calif.: RAND Corporation, RR-1954-A, 2019.

FIGURE 2.1
Logic of Multi-Period Assessment of Force Flow Model

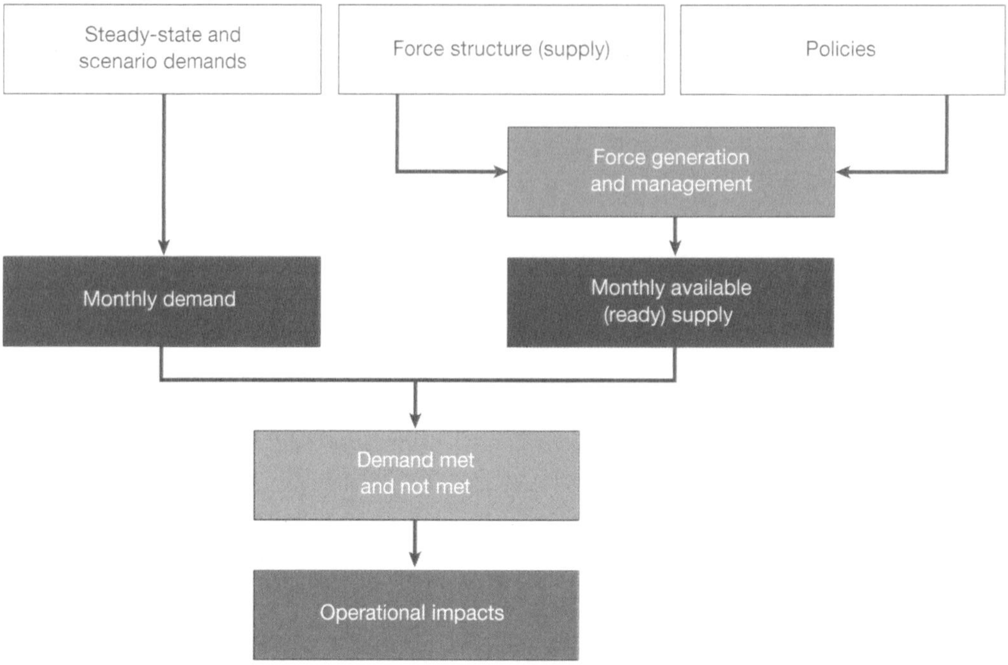

Steady-state and scenario demands
Force structure (supply)
Policies
Force generation and management
Monthly demand
Monthly available (ready) supply
Demand met and not met
Operational impacts

SOURCE: Best et al., 2019.

dation of Army force generation is the unit training cycle, which describes the amount of time a unit is expected to spend at a given readiness level for a given period. The Army's *Force Generation—Sustainable Readiness* publication, AR 525-29, illustrates the various training cycles that units may follow.[18] Within the unit training cycle, several parameters can be set to investigate how demand might be sourced against a given scenario. AR 525-29 specifies unit training cycles that apply to active and reserve component units. These differ primarily in the length of time spent in each readiness level; as one might anticipate, active component units achieve higher levels of readiness more quickly than their reserve counterparts. The model also has the flexibility to specify the initial readiness profile across all units to account for differences at a given point in time.

In filling rotational and emerging commitments, the user specifies allowable readiness levels for deployment, as well as priorities for whether the requirement will be filled by the active component or the reserve component. The user can specify the timing and length of deployments for a particular model run and the number of months of overlap that transitioning units should have for enduring commitments. The user can also specify alignment of units for particular geographic command requirements and units can be made unavailable

[18] ReARMM, forthcoming.

to support modernization efforts. Finally, the model allows the user to specify different D2D/M2D ratios for any excursion.

When one considers the variety of specifications for each of the model inputs used in describing unit availability, which are captured in Figure 2.2, the complexity quickly becomes apparent: There are thousands of possible combinations available for investigation. Therefore, it is important to note that it is not useful to view any output or set of outputs as optimal; rather, an output should be viewed as an observation of how unit demand is met or unmet for a particular investigation according to the set of assumptions and settings for the inputs that we have described. Naturally, any inference for a given model run must consider the set of input specifications.

Modeling Armored Brigade Combat Team Use in Competition

In practice, the demands placed on any force element vary over time. As an excursion to help establish an upper-boundary condition for what the inventory of ABCTs in the Army could possibly support during competition, demand over time is set to a constant number of requirements with no breaks in rotational demand. In doing this, one can quickly determine for a set of given input parameters the maximum number of rotational commitments the Army's ABCT inventory could plausibly support. For the set of parameters described, we changed the number of enduring commitments until there was a period of unmet demand exceeding

FIGURE 2.2
Inputs for Multi-Period Assessment of Force Flow Model

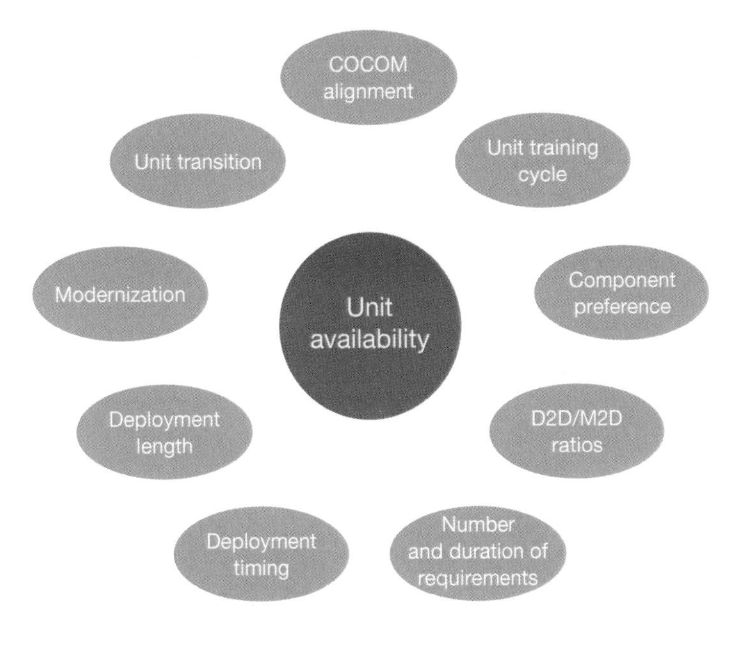

NOTE: COCOM = combatant command.

one month. Historically, these enduring rotational commitments have been between two and four ABCTs.[19]

We began with an evaluation of four enduring ABCT commitments that set the D2D/M2D ratios to their threshold levels (1:2 for the active component and 1:4 for the reserve component), which reflect a greater potential frequency, relative to the goal levels, for a unit to deploy. The 11 active component units follow a progressive C1, two-year cycle as described in AR 525–29, which requires that units maintain a C1 level of readiness for 50 percent of the time. The five reserve component units follow a progressive C2, five-year cycle as described in AR 525–29, which requires that units maintain a C2 level of readiness for 20 percent of the time. These training cycles were selected based on interviews with Army force managers because they reflect current training goals.

Additional input parameters for these model excursions require that 50 percent of the commitments are filled by active component units at a C1 level of readiness while the remaining 50 percent could be filled by either active or reserve component units at a C1 or C2 level of readiness. Each of the deployments is set to nine months in duration and requires a one-month overlap for transition between deployments (i.e., a one-month overlap for one unit to replace another unit). This nine-month deployment duration restricts reserve component preparation time for deployment to three months. The initial readiness profile (i.e., the number of units available by readiness level) reflects the unit training cycle goals of 50-percent C1 for the active component and 20-percent C2 for the reserve component. The four enduring commitments begin simultaneously for these excursions at month nine of the model run.[20]

As shown in the legend of Figure 2.3, the model captures *missed*, or unmet, demand; the numbers of active and reserve component units ready to fill commitments (and those less ready than preferred, if allowable); the total number of units demanded; other steady-state demand; and any reallocated steady-state demand. These last two entries are normally reserved for homeland security requirements.

For this group of input settings, demand for the four enduring commitments was fulfilled. As expected, given the requirement that 50 percent of these commitments be met by active component units at a C1 readiness level, the majority of the commitments were met by active component ABCTs, while the reserve component maintained suitable readiness and dwell to fulfill three deployments.

Changing the D2D/M2D policy from the threshold ratios to the goal ratios of 1:3 for the active component and 1:5 for the reserve component has significant impact on the ability of

[19] John R. Deni, *Rotational Deployments vs. Forward Stationing: How Can the Army Achieve Assurance and Deterrence Efficiently and Effectively?* Carlisle Barracks, Pa.: U.S. Army War College Press, August 25, 2017.

[20] Given that the purpose of this model was to evaluate the ability of Army ABCTs to meet simultaneous, enduring commitments, we initiated the demands for ABCTs at the same time. If the demands are initially separated in time, there are short-term differences in unit availability. But the model run length of 80 months captures the dynamic between readiness cycles and demand met or unmet in the long term for four enduring commitments. In other words, no matter how far apart one distributes the initial demands, the results we show occur once four commitments are in place.

FIGURE 2.3
Fulfilling Four Enduring Armored Brigade Combat Team Commitments at Threshold Deployment-to-Dwell/Mobilization-to-Dwell Ratios (1:2 for the active component, 1:4 for the reserve component)

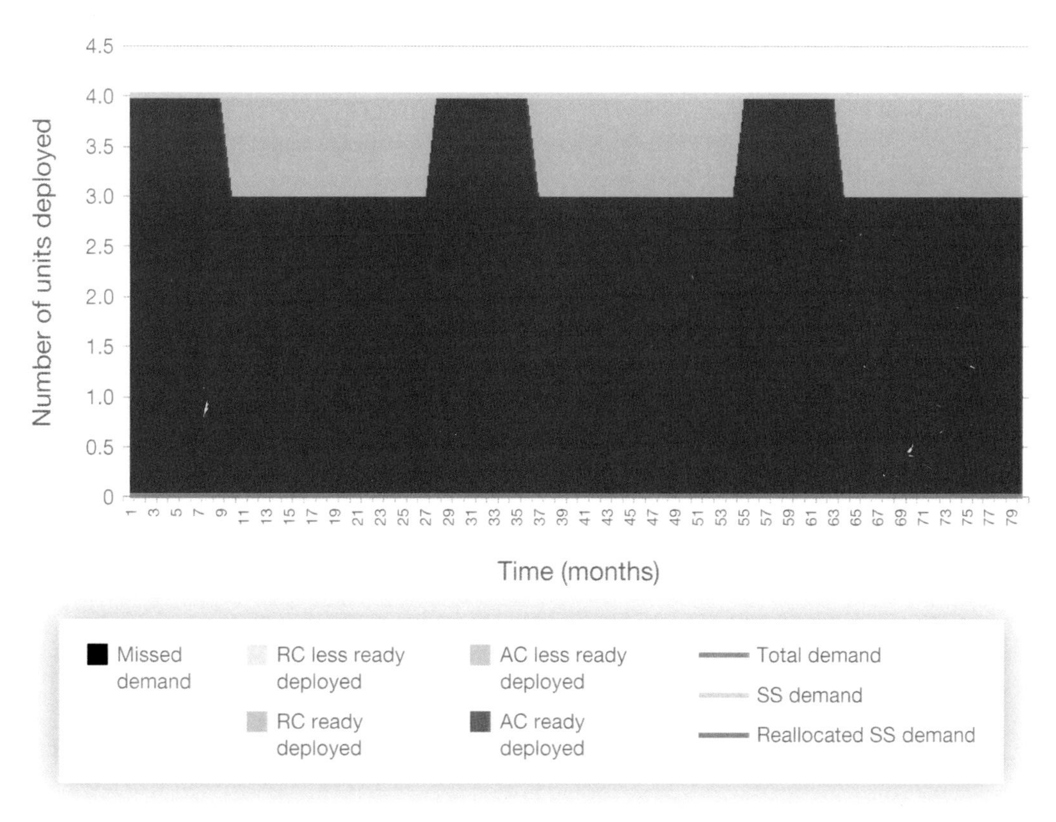

NOTES: AC = active component; RC = reserve component; SS = steady-state.

the ABCT inventory to meet the four enduring commitments for the input settings selected. Figure 2.4 illustrates the effect of the change in D2D/M2D ratios, which results in four periods of unmet unit demand (black areas) for the same time horizon when all other inputs are held constant. Although one of these periods is only three months (months 49–51), the other three periods of missed demand all exceed six months in duration; the longest of these is 13 months in duration (months 27–39).

Notably, the longest duration of unmet demand includes a significant period in which only 50 percent of demand (two of four requirements) was met; one of the two units fulfilled competition requirements, and it did so with deployed active component units that were less ready (months 31–39) than what the enduring commitment calls for. It is important to understand that this missed demand is reflective of a failure to meet a policy goal (e.g., a failure to meet demand through the use of nine-month rotations while maintaining D2D/M2D goal ratios) rather than an inability to *actually* meet the demand using a different policy choice.

FIGURE 2.4
Fulfilling Four Enduring Armored Brigade Combat Team Commitments at Goal Deployment-to-Dwell/Mobilization-to-Dwell Ratios (1:3 for the active component, 1:5 for the reserve component)

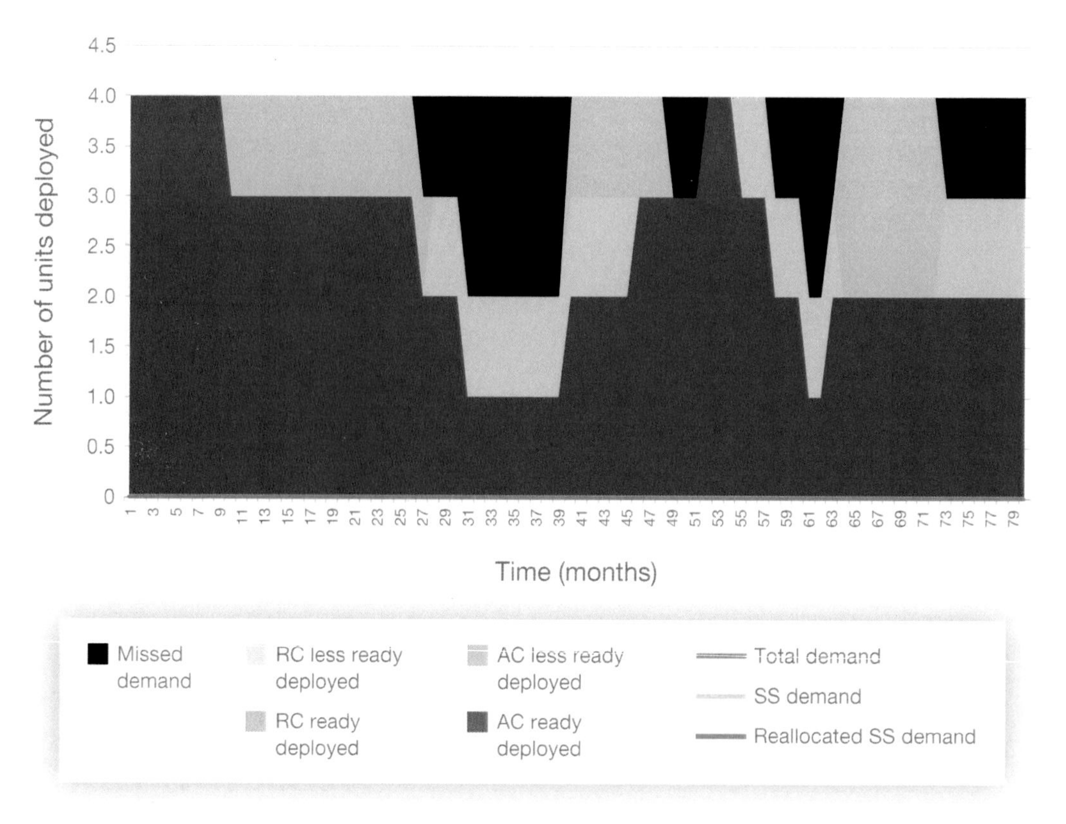

NOTES: AC = active component; RC = reserve component; SS = steady-state.

In addition to observing the impact of policy changes on meeting demand for ABCTs during a simulation run, it is important to understand the underlying readiness of units that were not deployed during this period. Figure 2.5 illustrates underlying readiness for the active and reserve components for both threshold and goal D2D/M2D ratios. The areas outlined purple compare underlying readiness for the active component for threshold (top panel) and goal (bottom panel) D2D ratios. While one might expect readiness to generally be better under goal ratios (more time at home), this simulation illustrates that this assumption is not inherently true throughout any given period.

The areas outlined purple show a period in which, on average, underlying readiness for ABCTs operating under the goal ratio (bottom panel) for the active component was less than that enjoyed by ABCTs operating under the threshold ratio (top panel). These readiness results could be attributed to any number of things (e.g., changing the initial readiness profile selected for the simulation changes the resulting readiness profiles), but the overarching

FIGURE 2.5

Underlying Readiness for Undeployed Units During Simulation

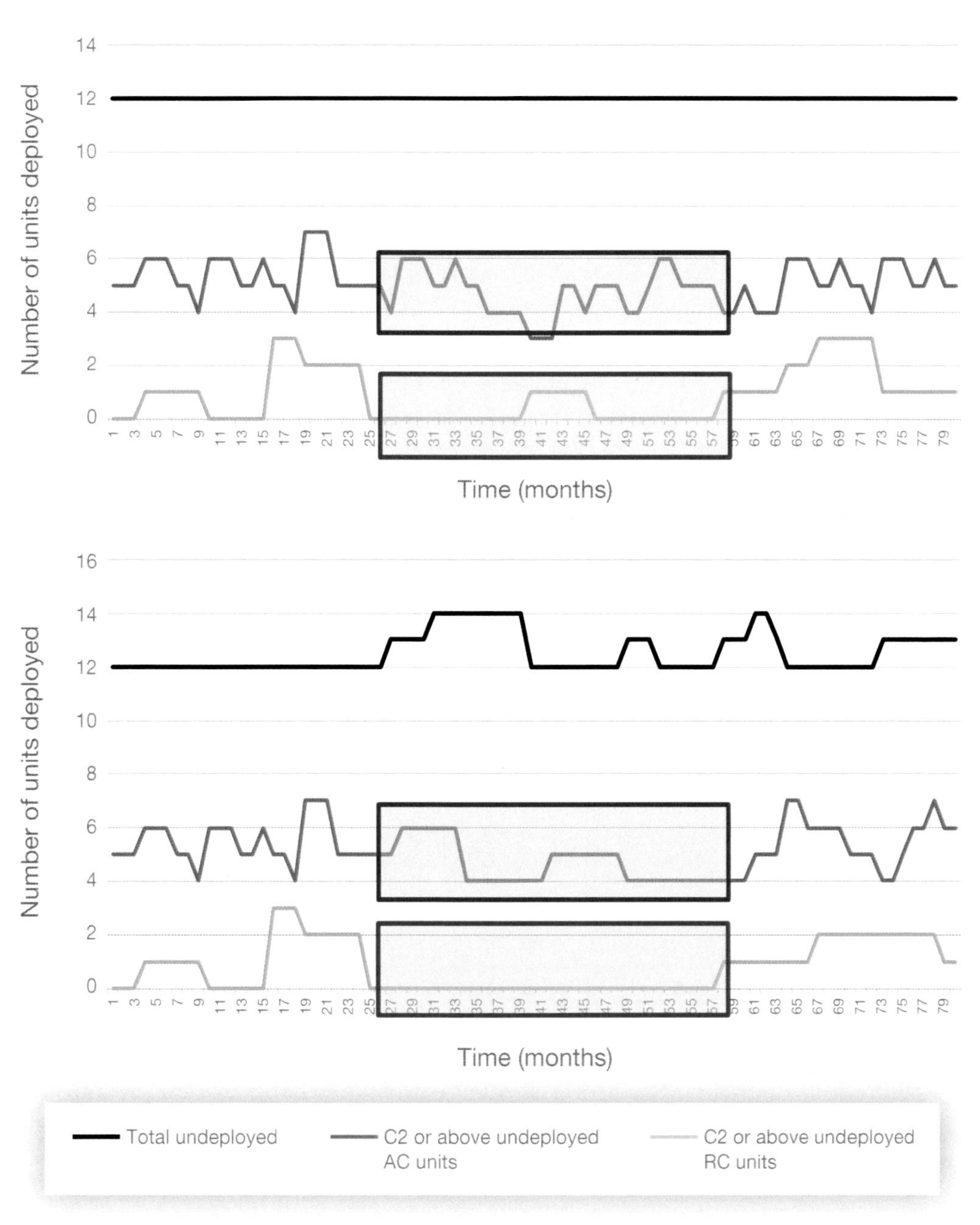

NOTES: AC = active component; RC = reserve component. The top panel shows underlying readiness under threshold D2D/M2D ratios (1:2 for the active component, 1:4 for the reserve component); the bottom panel shows underlying readiness under goal D2D/M2D ratios (1:3 for the active component, 1:5 for the reserve component). The areas outlined purple compare underlying readiness for the active component for threshold and goal D2D ratios. The areas outlined blue compare underlying readiness for the reserve component for threshold and goal M2D ratios.

point is that there is no guarantee that unit-level readiness will be significantly higher under the goal ratio.

The unit training cycle is carefully defined and operates independently of D2D/M2D policy. For the competition period examined, with four enduring commitments, there were only small differences in the active component overall, in which units were at C2 or better readiness levels. In total, active component readiness was at C2 or above for 405 unit-months under the threshold ratio, compared with 409 unit-months under the goal ratio. These overall readiness ratings likely stand in contrast to the other components of readiness, such as personnel or equipment readiness, which may very well benefit from greater time at home.

The areas outlined blue in the figure compare underlying readiness for reserve component ABCTs for threshold (top panel) and goal (bottom panel) M2D ratios. For modeling unit availability using this set of input parameters, no reserve component units were available to deploy at C2 readiness or better (months 26–59) for an extended period. As with the active component, the overall numbers of unit-months of availability at C2 readiness or better were similar for the two M2D ratios for the reserve component. Under a threshold ratio of 1:4, there were 71 unit-months of undeployed availability at C2 readiness or better for reserve ABCTs, as compared with 62 unit-months of undeployed availability at C2 readiness or better under the goal ratio of 1:5. The difference between the two occurs because of an additional deployment fulfilled by the reserve component when operating under the goal ratio.

The analysis of ABCTs fulfilling four enduring commitments during competition provides a useful case study that one can use to determine the largest number of commitments that this JFE is capable of fulfilling when operating under goal or threshold D2D/M2D ratios for a particular set of input parameters. There are some parameter settings that should be brought to the reader's attention. Restricting D2D ratios to round numbers results in diminished unit availability. A nine-month rotational commitment requires the active component units to fulfill under a D2D ratio of 1:2. Changing the threshold ratio for the active component from 1:2 to 1:1.67 would reduce the number of units required from four to three but would result in a reduction in dwell from 18 months to 15 months.

Similar small changes to the ratios might be possible for the reserve component, but the benefits would take a significantly longer time to be realized given the relatively small amount of time spent at C2 readiness levels. Meeting the Army's goal of a 1:3 D2D ratio would require 4.44 (in practical terms, five) active component units. In addition, this analysis assumes a one-month transition period between units. Reducing this transition requirement would achieve similar results to changing the D2D ratio, resulting in one less unit needed to fulfill a commitment but with no accompanying requirement to change the D2D ratio.

Summary

D2D/M2D policy is a principal driver for how rotational commitments are managed for the Army. Interviews with individuals across the Army's force management community indicated that force managers continuously balance unit availability based on future readiness profiles with required demands and forecasted D2D/M2D ratios. If the primary purpose of the policy is to be a management tool for rotational commitments so that commitments for service members and employers are more predictable, the policy seems to be achieving its purpose for ABCTs. Beyond near-term force management decisions, the Army uses D2D/M2D ratios in analysis that informs future force structure decisions.

We observe that current threshold and goal D2D/M2D ratios are independent of Army readiness objectives. The Army generates ready units on a cycle either faster than (for the active component) or at the same rate as (for the reserve component) current threshold D2D/M2D ratios. This force generation policy creates a pool of ready units that, because of D2D/M2D constraints, are not available for—and thus cannot be used for—planned rotational deployments. However, the pool would be ready for an unplanned contingency, when D2D/M2D constraints would not apply. Of course, the size of such a pool would vary with actual readiness-generation performance and ongoing and recently completed deployments.

Our modeling of ABCT employment illustrated that the differences between goal and threshold D2D/M2D ratios have little to no impact on unit availability for both the active and reserve components under the scenarios examined. Notably, this modeling is based on the Army's Sustainable Readiness Model. As of the writing of this report, the Army is transitioning to another force generation model that might yield different results from those shown in this analysis.

Finally, we found that very small changes in directed ratios can alleviate stress on the number of units required to fulfill a commitment while maintaining the same level of readiness. However, such a policy change would result in less time spent in dwell status for service members.

CHAPTER THREE

Deployment- and Mobilization-to-Dwell Policy in the U.S. Marine Corps

In this chapter, we examine D2D/M2D policy in the Marine Corps. Analysis of the selected JFE, the infantry battalion, is based on the same approach used for the ABCT in the previous chapter. We describe adaptations to the MPAFF model to capture differences in how the Marine Corps trains and deploys compared with the Army, but the reader should refer to the previous chapter for a full discussion of the MPAFF tool. We begin this chapter with a description of how the policy manifests in the Marine Corps. Following this description, we present several model excursions for the infantry battalion focusing on the impacts of D2D/M2D policy during competition.

Policy

The Marine Corps has a published D2D/M2D policy that is separate from but closely aligned with the OSD policy. Marine Corps Administrative Message (MARADMIN) 346/14, "Deployment-to-Dwell, Mobilization-to-Dwell Policy Revision," published in 2014, describes how the Marine Corps will consider and comply with D2D/M2D ratios.[1] MARADMIN 346/14 is based on the 2013 Secretary of Defense D2D/M2D memorandum and clearly states the purpose of the policy; defines *operational deployment*, *dwell*, and *ratios*; and identifies goal and threshold ratios that are aligned with DoD ratios.

Per the MARADMIN, as in the OSD policy, the threshold, or redline, D2D/M2D ratios are 1:1 and 1:4 for the active component and the reserve component, respectively. Similarly, the goal ratios are 1:2 and 1:5. To deploy a unit that does not meet the threshold for dwell or demobilization requires a Secretary of Defense–signed waiver, which the MARADMIN spells out. As per OSD policy, an active component unit is considered deployed when the majority of the unit is engaged in a Secretary of Defense–approved operational requirement away from home station. A reserve component unit is considered mobilized per the terms of Secretary of Defense–approved orders so stating.

---

1 MARADMIN 346/14, "Deployment-to-Dwell, Mobilization-to-Dwell Policy Revision," Washington, D.C.: U.S. Marine Corps, July 14, 2014.

The MARADMIN goes a step further than the 2013 Secretary of Defense D2D/M2D memorandum by requiring the assessment of individual marines' D2D/M2D ratios within units and as individual augmentees (IAs) for active component and reserve component units prior to operational deployments. The Marine Corps tracks D2D and M2D at the unit level and higher, and the MARADMIN spells out processes for unit compliance or waivers and for individual compliance or waivers.

Similar to what was observed in DoD writ large and across the different services, since 2016, the Marine Corps has issued more-recent guidance stating the goal of returning to 1:3 D2D and 1:5 M2D and moving away from the threshold ratios. Testifying before the U.S. Senate Armed Services Committee Subcommittee on Readiness in March 2016, then–Assistant Commandant of the Marine Corps Gen John Paxton stated that the Marine Corps' ideal ratio for D2D is 1:3, noting that

> home station time is required for the unit to conduct personnel turnover, equipment reset and maintenance, and complete a comprehensive individual, collective, and unit training program across all their mission essential tasks (METs) prior to deploying again.[2]

In March 2020, Marine Corps Commandant Gen David H. Berger stated in his testimony before the Senate Armed Services Committee that the Marine Corps "will continue to field an elite Active and Reserve Marine force, maintaining a 1:2 deployment to dwell ratio while working towards a necessary 1:3 ratio to preserve constant readiness and availability of personnel while also reserving time for training, refitting and family support."[3] Although this statement implies that a move to a 1:3 ratio is necessary to preserve readiness, the Marine Corps also acknowledges that some deployments increase readiness.[4] Finally, a November 2020 report by the Heritage Foundation echoes these remarks, noting that the Marine Corps is sustaining a 1:2 D2D ratio while moving in the direction of a 1:3 ratio.[5]

---

[2] John Paxton, Assistant Commandant of the Marine Corps, "U.S. Marine Corps Readiness," statement before the Subcommittee on Readiness, Committee on Armed Services, U.S. Senate, Washington, D.C., March 15, 2016, p. 8.

[3] Thomas B. Modly, Acting Secretary of the Navy; Michael M. Gilday, Chief of Naval Operations; and David H. Berger, Commandant of the U.S. Marine Corps, "Fiscal Year 2021 Department of the Navy Budget," statement before the Senate Armed Services Committee, U.S. Senate, Washington, D.C., March 5, 2020, p. 18.

[4] U.S. Marine Corps Forces, Europe and Africa, "U.S. Marines and Sailors from Marine Rotational Force-Europe Complete Deployment to Northern Norway, Return to Camp Lejeune, N.C.," Defense Visual Information Distribution Service, April 16, 2021.

[5] Dakota L. Wood, "An Assessment of U.S. Military Power: U.S. Marine Corps," Heritage Foundation, November 17, 2020.

Deployment-to-Dwell/Mobilization-to-Dwell Management

As the Marine Corps policy (which aligns with the OSD policy) demands, D2D/M2D status provides the service with a key piece of information through which to assess unit availability to respond to enduring and emerging requirements. Specifically for the Marine Corps, the policy also provides the service with information on individual availability to deploy.

As with the Army requirements, the Marine Corps requirements are dictated through the GFMAP. For the Marine Corps, Marine Forces Command (MARFORCOM) is principally responsible for coordinating and recommending GFMAP sourcing. In so doing, MARFORCOM considers the current and projected D2D/M2D statuses of units to determine their availability to meet given requirements. The D2D/M2D statuses of units chosen to source a requirement are logged in Annex A of the GFMAP, a continuously updated record of requirements and sourcing decisions.

D2D and M2D are also used by MARFORCOM to manage unforeseen requirements not captured in the GFMAP process. When confronted with such a demand, MARFORCOM planners consider the D2D/M2D statuses of units that can provide the required capability, whether that capability can be delivered through normal structures, and whether they have the authority to use the reserve component. MARFORCOM planners report that D2D and M2D are considered but do not drive deployment schedules.[6]

Thus, the Marine Corps tracks individual D2D and M2D along with PERSTEMPO events—those events that do not rise to the threshold of an operational deployment but that keep marines from home. These elements are tracked at the unit level and higher and can be accessed within each unit to assess individual D2D and M2D.[7]

The Marine Corps Infantry Battalion

For the Marine Corps, we analyzed the infantry battalion, a basic building block of Marine Corps ground combat power. The mission of the Marine Corps infantry battalion is to "locate, close with, and destroy the enemy by fire and maneuver or to repel an enemy's assault by fire and close combat."[8] An infantry battalion is composed of approximately 900 personnel spread across five companies—a headquarters company, a weapons company, and three rifle companies. The rifle company is the primary tactical unit.[9]

The Marine Corps contains 24 active component battalions located across the United States—nine battalions on Camp Lejeune, North Carolina; 12 battalions between Camp

[6] Telephone discussion, Marine Forces Command, February 23, 2021.

[7] Telephone discussion, Marine Forces Command, February 23, 2021.

[8] Marine Corps Reference Publication (MCRP) 1-10.1, *Organization of the United States Marine Corps*, Washington, D.C.: U.S. Marine Corps, August 26, 2015, p. 5-5.

[9] MCRP 1-10.1, 2015, p. 5-5.

Pendleton and Twentynine Palms, California; and three battalions at Kaneohe Bay, Hawaii—and eight reserve component battalions at various locations. The Marine Corps' *Force Design 2030* proposes a transformation of how the institution is organized, trained, and equipped, which includes reducing infantry battalions and associated support. At the time of this writing, the specifics of that transformation have not been published.[10]

The Marine Corps' infantry battalions face a variety of ongoing operational commitments. These commitments include three Marine expeditionary units (MEUs): one that traditionally deploys to the Mediterranean/Middle East region, primarily from Camp Lejeune; one that deploys to the U.S. Indo-Pacific Command region, primarily from Camp Pendleton; and one that deploys throughout parts of Southeast Asia, based in Okinawa. The Marine Corps also sends three battalions at a time to Okinawa to participate in the Unit Deployment Program. Finally, the Marine Corps deploys a battalion each to the Special Purpose Marine Air Ground Task Force–Crisis Response–Africa and Central Command; a battalion to Darwin, Australia, as part of the Marine Rotational Force-Darwin; and a subordinate element of a battalion to Norway as part of the Marine Rotational Force-Europe. Although the Marine Corps states that it globally sources its commitments, those sourcing solutions tend to be predictable, with the East Coast (Camp Lejeune) battalions primarily sourcing the Mediterranean/U.S. Central Command MEU and the West Coast (Camp Pendleton/Twentynine Palms) battalions primarily sourcing the U.S. Indo-Pacific Command MEUs.

Modeling Marine Corps Infantry Battalion Availability

We used the MPAFF model described in Chapter Two to analyze the impacts of D2D/M2D policy on the Marine Corps infantry battalion. In this section, we describe how we used the model to properly capture Marine Corps deployments given that differences exist between Army and Marine Corps training and unit deployments in policy and practice.

Key Model Inputs and Assumptions

As discussed in Chapter Two, the unit training cycle is a key input to the MPAFF model. In general, Army units follow a predetermined cycle designed to generate readiness: Units move through different stages of readiness (i.e., C-levels) over time, and a unit can deploy only when it is at the required level of readiness. Absent D2D/M2D requirements, this can, in some cases, lead to units being unable to deploy because they have transitioned out of peak readiness and into a lower state of readiness.

Marine Corps infantry battalions, by contrast, do not follow a fixed training and readiness cycle and instead generate readiness according to deployment requirements and time-

[10] For more information, see U.S. Marine Corps, *Force Design 2030*, Washington, D.C., March 2020; and U.S. Marine Corps, *Force Design 2030: Annual Update*, Washington, D.C., April 2021.

lines. Thus, we do not see units moving in and out of "ready" or "deployable" status based on their readiness ratings. Instead, units are "ready" and "deployable" when deployed and "not ready" when not deployed. The D2D/M2D ratios seem to be the driving factors behind these periods, with units spending the time between deployments generating readiness at an appropriate rate to be "ready" when the next deployment occurs. To capture this nuance within the model, we do not assume any readiness-generation cycle through which units move and, instead, allow the D2D/M2D policy to govern the statuses of units. That is, units are considered to always be at the required state of readiness and are nondeployable only for a period immediately following a deployment, as dictated by the D2D/M2D policy.

While the MPAFF model allows for variance in the timing and length of deployments across different requirements, historical infantry battalion deployment data have suggested that deployments tend to last six months, with most occurring on a heel-to-toe basis. Thus, our modeling assumes these parameters. The main exception is the 31st MEU, which deploys annually from Japan. The two MEUs deploying from the United States tend to exhibit gaps between consecutive deployments, and the average gap of four months is reflected in our modeling. Finally, although users may specify the alignment of units across specific combatant commands within MPAFF, we assumed global sourcing of Marine Corps infantry battalions for all requirements. The MPAFF inputs shown in Figure 2.2 also play a key role in Marine Corps infantry battalion modeling.

Modeling Marine Corps Infantry Battalion Use in Competition

Enduring Requirements

As discussed previously, Marine Corps infantry battalions have approximately eight enduring requirements, with between six and eight occurring simultaneously at any given time. As with the Army ABCT analysis, we begin with an evaluation that sets the D2D/M2D ratios to their threshold levels (1:2 for the active component and 1:4 for the reserve component), with each deployment lasting six months. As noted previously, infantry battalions generate readiness according to deployment requirements, so, within the model, active component units are considered ready to deploy at all times except the 12-month period immediately following a deployment. (A six-month deployment and D2D ratio of 1:2 requires a 12-month dwell period.) Similarly, reserve component units are considered ready at all times except the 48-month period immediately following a deployment. (Note that reserve component units are mobilized for one year—six months of mobilization and predeployment training followed by a six-month deployment—which requires four years of dwell.)

Our analysis of enduring requirements indicates that the active component is able to fulfill all eight enduring commitments without a need for the reserve component, as illustrated in Figure 3.1. Intuitively, this makes sense because a D2D ratio of 1:2 implies that three units are required for every commitment (assuming heel-to-toe deployments) and that the 24 active component infantry battalions are sufficient to cover these requirements. Absent additional requirements (e.g., surge deployments), use of the reserve component would be necessary

FIGURE 3.1
Fulfilling Eight Enduring Infantry Battalion Commitments at Threshold Ratios (1:2 for the active component, 1:4 for the reserve component)

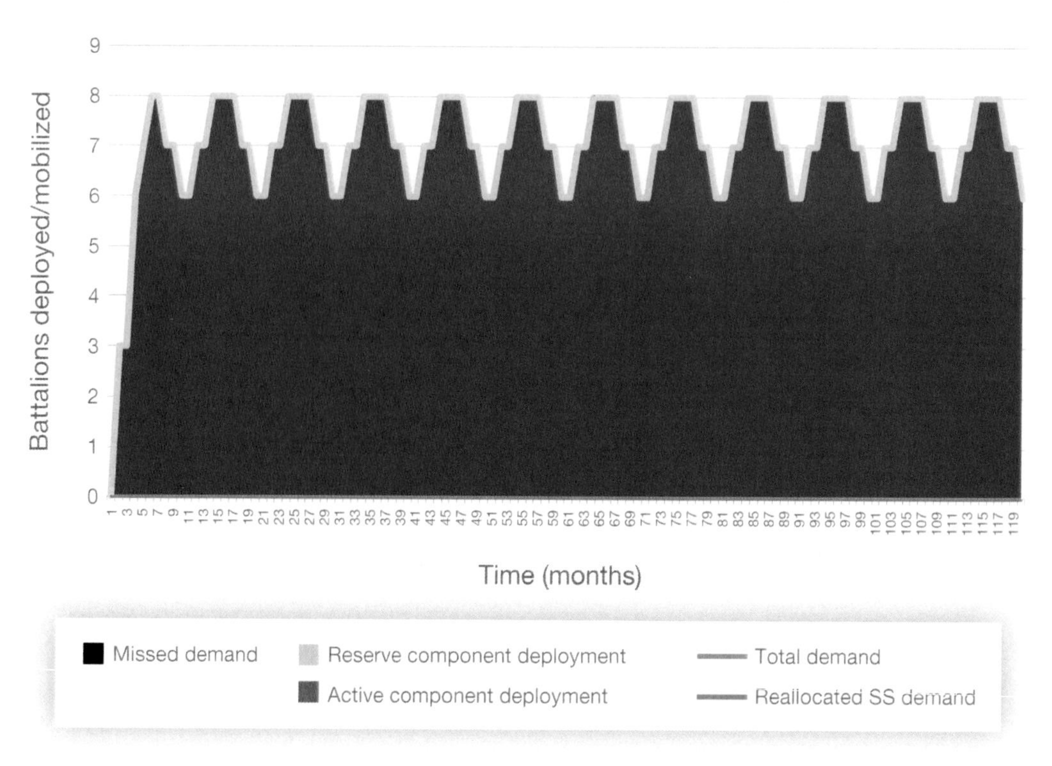

NOTE: SS = steady-state.

only in cases in which an active component deployment exceeded the typical six months.[11] Thus, it seems that the Marine Corps is scheduling operational deployments to maximize use of the active component.

Changing D2D/M2D ratios from threshold to goal ratios of 1:3 for the active component and 1:5 for the reserve component, however, impedes the ability of infantry battalions to meet all requirements. As Figure 3.2 shows, full use of the reserve component is required (i.e., all eight reserve units must be utilized every six years), yet approximately 9 percent of demand remains unfulfilled. These periods of unmet demand vary in length from as short as two months to as long as 11 months (months 47–57), and the number of unmet demands at any given time also varies.[12] During some periods, as many as five requirements cannot be met because of insufficient inventory (under the assumed conditions). To fulfill goal ratios,

[11] Lack of need for the reserve component to fulfill the current set of enduring commitments aligns with historical data that show that reserve component units were called upon only twice from FY 2015 to FY 2021.

[12] Similar to the Army model results, this unmet demand is not the inability to actually meet demand but the implication of the policy choice under assessment. A different policy choice might yield different results.

FIGURE 3.2
Fulfilling Eight Enduring Infantry Battalion Commitments at Goal Ratios (1:3 for the active component, 1:5 for the reserve component)

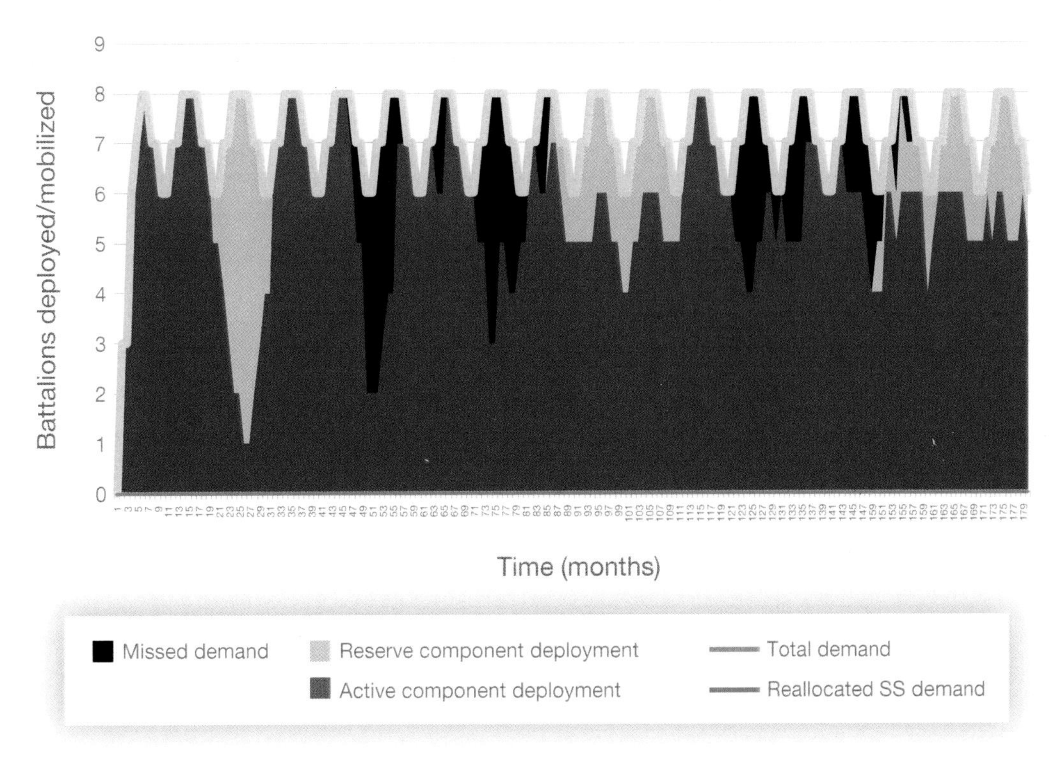

NOTE: SS = steady-state.

the Marine Corps would need to reduce operational deployments, even with the use of the reserve component.

We noted earlier that the Marine Corps appears to be scheduling (current) enduring deployments to maximize active component utilization, and the immediately preceding analysis demonstrates that transitioning to goal ratios will challenge its ability to meet these requirements in the future. Therefore, for several potential future scenarios that Marine Corps infantry battalions may encounter, we assess the impacts of D2D/M2D policy on the ability of the infantry battalions to meet demand.

Excursion 1: Reduction in Requirements

We first consider a scenario in which infantry battalions experience a reduction in enduring commitments while operating at goal D2D/M2D ratios. Specifically, we assume that the Unit Deployment Program requirement has been reduced from three battalions to two battalions. This assumption results in a requirement of between five and seven battalions at any given time. As shown in Figure 3.3, these requirements can be nearly 100 percent fulfilled (with the use of the reserve component) when the battalions are operating at goal D2D/M2D ratios. In

FIGURE 3.3
Fulfilling Seven Enduring Infantry Battalion Commitments at Goal Ratios (1:3 for the active component, 1:5 for the reserve component)

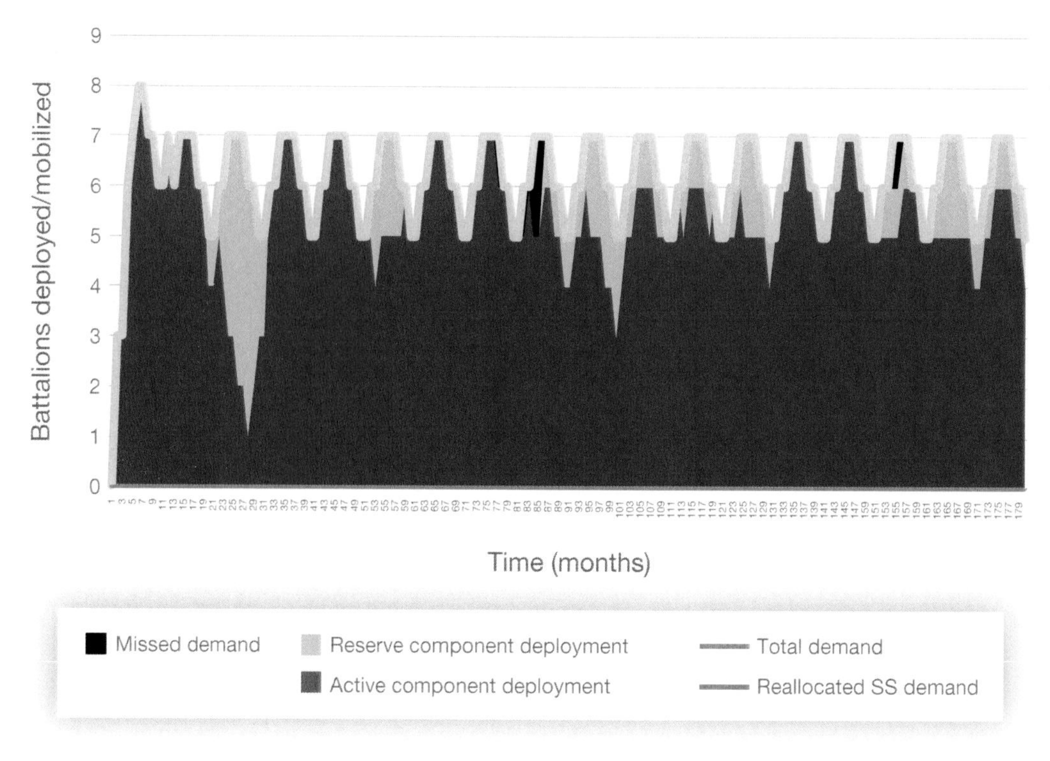

NOTE: SS = steady-state.

the long run, Marine Corps infantry battalions will be unable to meet approximately 1 percent of demand under the assumed conditions. However, to achieve this high rate of satisfaction under goal ratios, even with reduced demand, requires regular use of the reserve component. Approximately 11 percent of demand is satisfied by reserve units.

Excursion 2: Increase in Requirements

Next, we consider a situation in which Marine Corps infantry battalions experience an increase in enduring commitments, from eight to nine requirements. (We assume that the added requirement is in the form of a heel-to-toe deployment that begins in month 13.) Figure 3.4 shows the fulfillment, or lack of fulfillment, of these requirements under both threshold (top panel) and goal (bottom panel) ratios. It is evident that the fulfillment of all nine requirements is feasible when battalions are operating at threshold ratios with limited yet steady use of the reserve component. The active component can fulfill 96 percent of demand, and the reserve component fulfills the remaining 4 percent.

At goal ratios, however, the infantry battalions can no longer fulfill the increase in requirements. For eight enduring commitments at goal ratios, 9 percent of demand is unfilled; adding

FIGURE 3.4
Fulfilling Nine Enduring Infantry Battalion Commitments

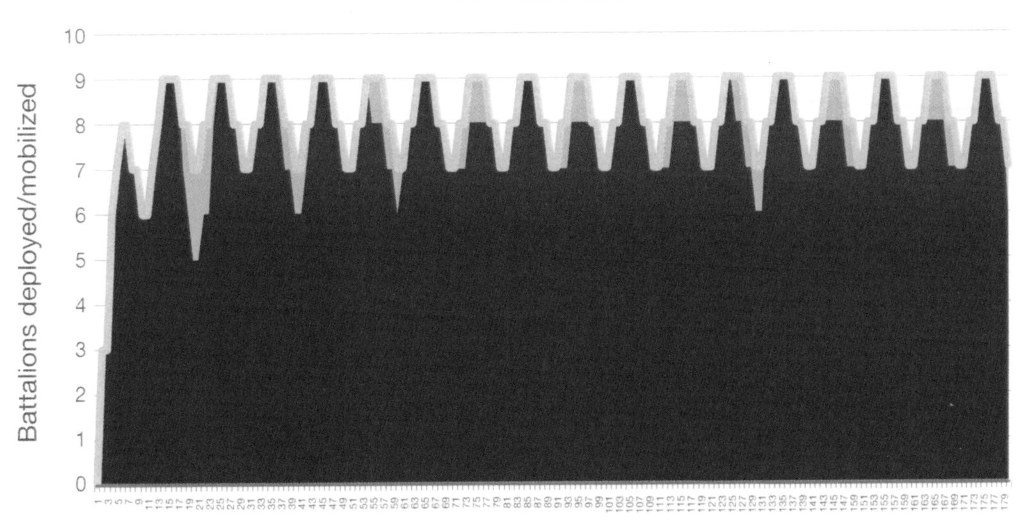

Goal ratios

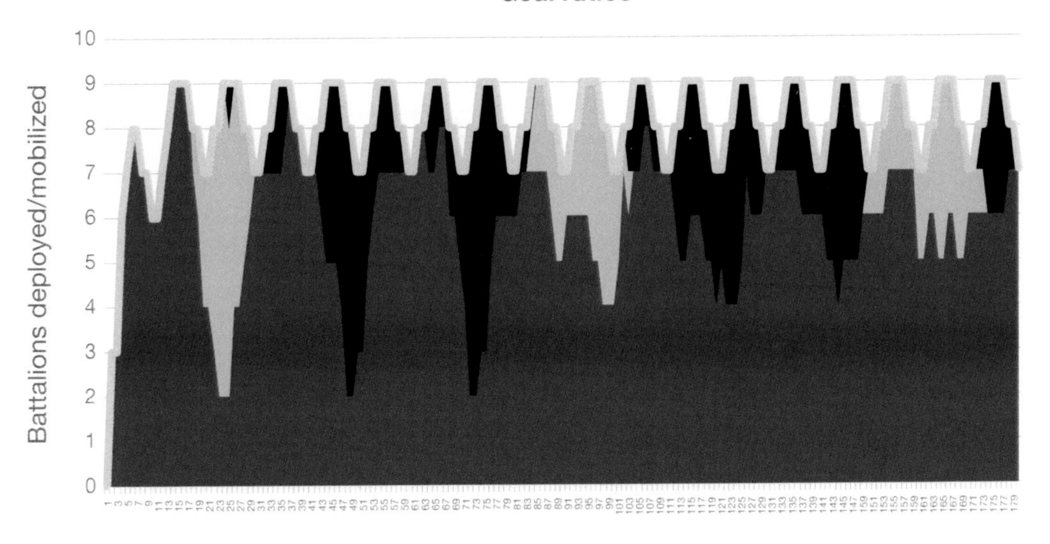

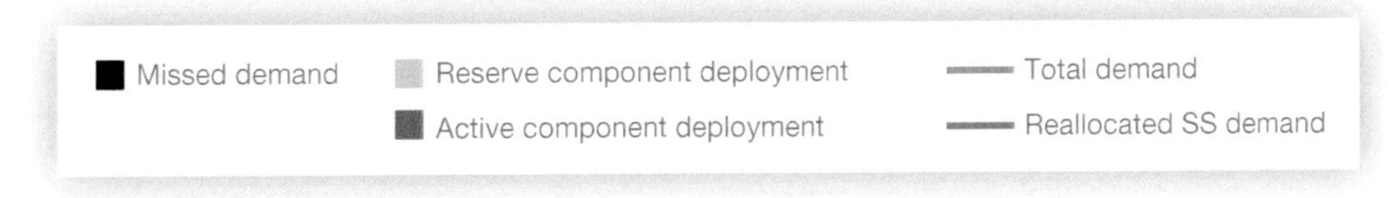

NOTE: SS = steady-state.

a ninth deployment requirement nearly doubles the amount of unmet demand. In this case, 17 percent of demand goes unmet, the active component fulfills 73 percent, and the reserve component fulfills 10 percent. Therefore, to meet an increase in demand would require both adherence to threshold ratios and the use of the reserve component.

Excursion 3: Reduction in Active Component Units

Given that a reduction in infantry battalion inventory is a consideration in *Force Design 2030*, we consider situations in which the active component inventory has been reduced by one and two battalions when both the active and reserve components are operating at goal D2D/M2D ratios. The top chart of Figure 3.5 shows that Marine Corps infantry battalions can continue to fulfill all eight enduring commitments should the active component experience an inventory reduction of one battalion. However, limited yet steady use of the reserve component is required once again and fulfills approximately 4 percent of demand. A reduction in active component inventory by two battalions, however, significantly increases reserve component utilization—11 percent of demand fulfilled by the reserve component—and results in approximately 14 percent of demand unmet. Compared with the current inventory, however, this is a relatively modest increase in unmet demand, as the current inventory results in approximately 11-percent missed demand under goal ratios.

Excursion 4: Active Component at Goal and Reserve Component at Threshold

In our final excursion, we consider the impacts of the active component operating at the goal D2D ratio (1:3) and the reserve component operating at the threshold M2D ratio (1:4). The top chart of Figure 3.6 shows the fulfillment of demand when both the active and reserve components operate under goal ratios. The bottom chart shows how demand is filled if the reserve component instead continues to operate under its threshold ratio of 1:4. Here, we see the expected earlier use of the reserve component during its second wave of utilization. However, there is little impact on the overall ability of the Marine Corps infantry battalions to fulfill their demands. Approximately 9 percent of demand still cannot be met.

To summarize our modeling observations, our analysis, based on historical data, shows that current operational requirements limit the flexibility that the Marine Corps might need to respond to unexpected crises or to modernization demands given the D2D/M2D policy, leadership guidance, and the threshold and goal ratios.

Our analysis shows that active component battalions can sustain eight ongoing deployments at threshold ratios of 1:2 but can fill only 80 percent of ongoing deployment requirements at goal ratios of 1:3. If reserve component battalions are used to bridge the gaps under goal ratios, an additional 11 percent of the requirements are met, leaving 9 percent unmet. Using the reserve component and reducing requirements to seven ongoing deployments leaves 1 percent of demand unmet. If ongoing deployments are reduced to meet goal ratios, then reducing the number of active component battalions might be feasible for the Marine Corps.

FIGURE 3.5

Fulfilling Enduring Infantry Battalion Commitments with Reduced Active Component Inventory and Goal Ratios (1:3 for the active component, 1:5 for the reserve component)

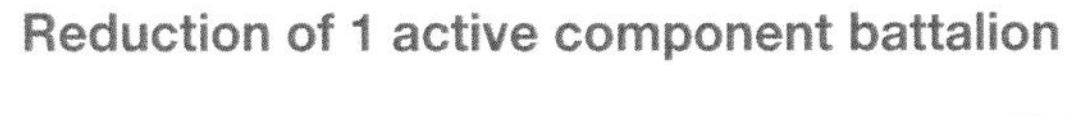

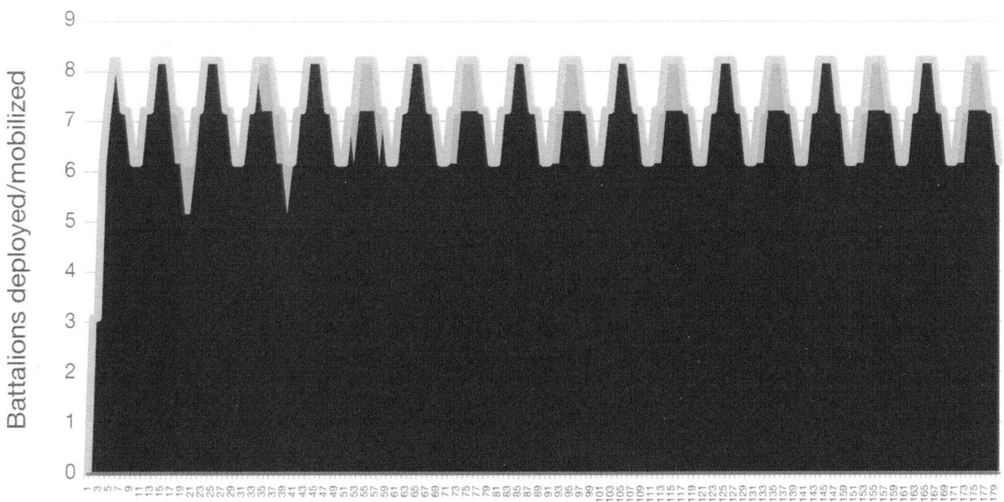

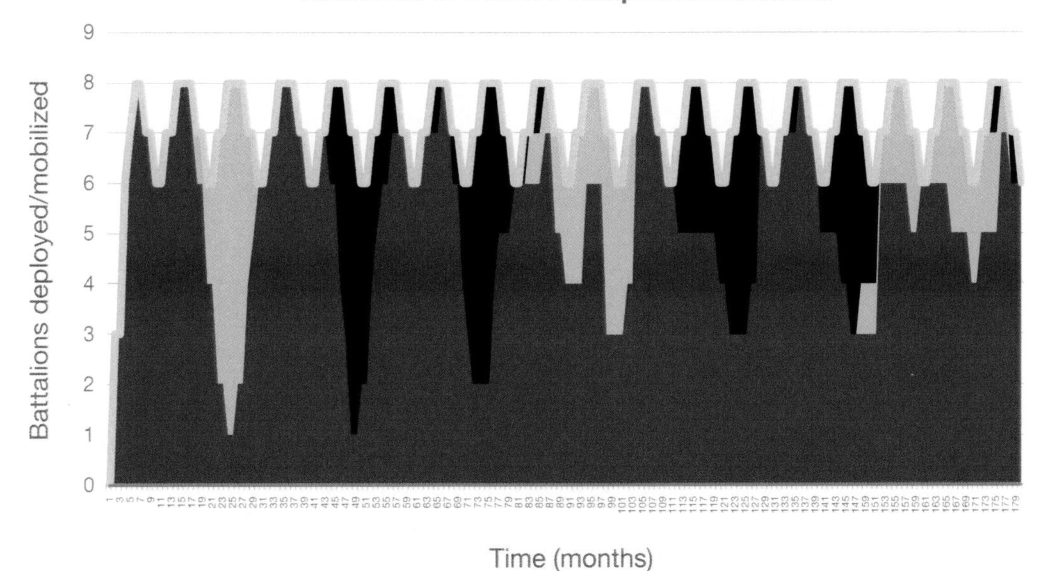

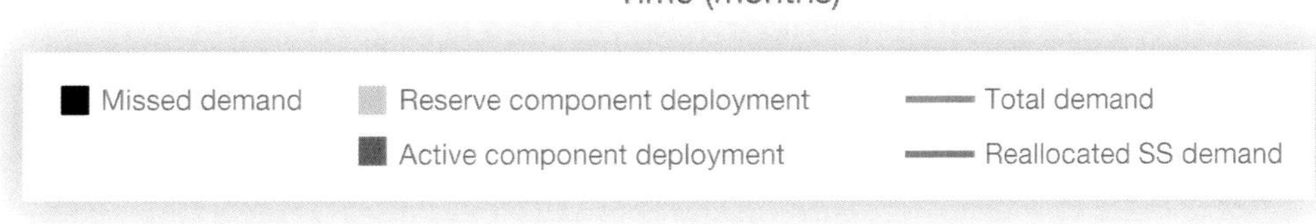

NOTE: SS = steady-state.

FIGURE 3.6

Fulfilling Eight Enduring Infantry Battalion Commitments with Active Component at Goal Ratio and Reserve Component at Threshold Ratio (1:3 for the active component, 1:4 for the reserve component)

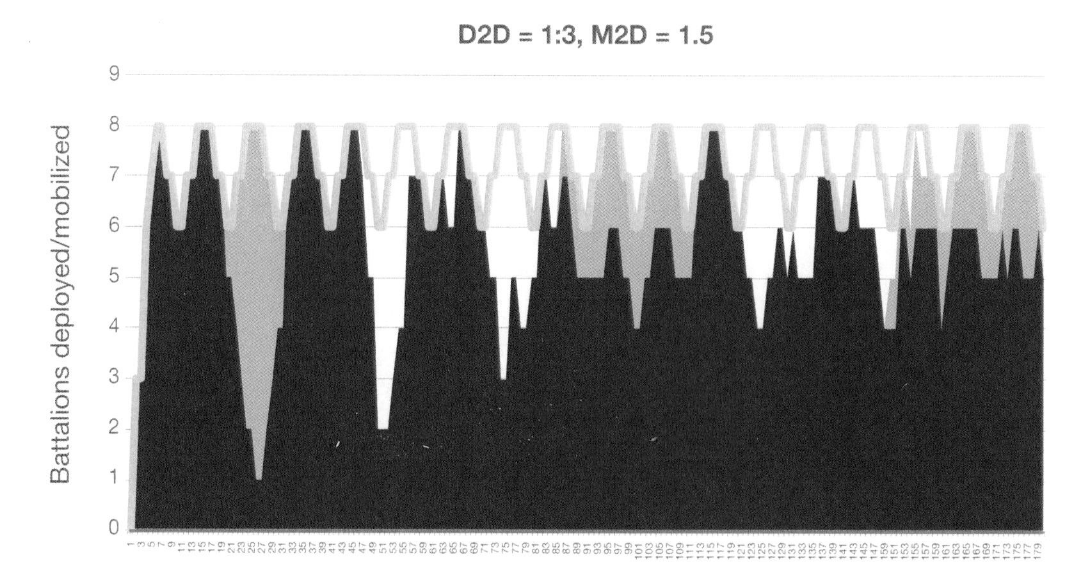

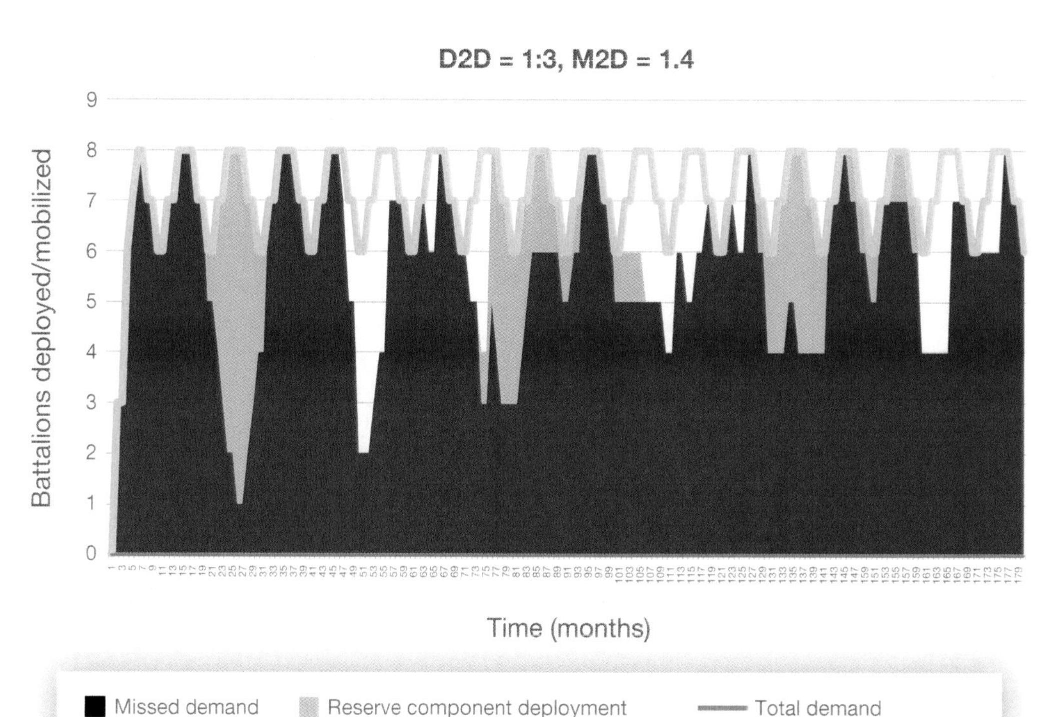

NOTE: SS = steady-state.

Summary

Findings from our analysis suggest that the D2D/M2D policy plays a central role in force management decisions but that the specific ratio for the reserve component, whether goal or threshold, does not have much effect on missed demand. From our analysis of the Marine Corps infantry battalion, historical data show that the Marine Corps schedules operational deployments to maximize active component utilization under the threshold ratio. Reserve component use remains a solution when active component deployments exceed the typical six months and are not a consistent part of scheduling for enduring requirements. Given that the Marine Corps Commandant's *Force Design 2030* calls for goal ratios along with a reduction in the number of infantry battalions, questions remain about how these objectives can be accomplished.[13] Could Unit Deployment Program rotations also cover additional requirements? Could the Marine Corps potentially present or structure forces in different numbers or ways? Because the Marine Corps actively tracks individual D2D/M2D, could this information inform force structure and deployment rotations in different ways?

Divesting force structure for modernization or meeting increasing requirements calls for additional considerations in the context of the D2D/M2D policy and associated guidance. These considerations include assessing demand to determine requirement flexibility; assessing the impacts of changes to numbers, structures, and the mix of active and reserve component units; and exploring alternative means of training—that do not involve deployment—to ensure readiness.

[13] U.S. Marine Corps, 2020.

CHAPTER FOUR

Deployment- and Mobilization-to-Dwell Policy in the U.S. Air Force

In examining USAF implementation of D2D/M2D policy and its effects on meeting competition requirements for forces, the sponsor selected for its case study the KC-135 Stratotanker aerial refueling capability, which is a USAF high-interest JFE. This assessment is restricted to a notional steady-state deployment demand to illustrate how the D2D/M2D policy could manifest in the future for this force element and implications that the policy could have on readiness profiles of the active and reserve components over time for a specified period of competition. The modeling results also provide a means for decisionmakers to understand the availability of the KC-135 to support day-to-day activities outside GFMAP via taskings from U.S. Transportation Command (USTRANSCOM). This chapter contains the results of this case study analysis and includes discussion of how the policy is implemented and managed in the USAF and the methodology used in conducting this assessment.

Policy and Data

The USAF uses the D2D/M2D ratios established in OSD's policy as primary metrics for reporting OPTEMPO to internal and external audiences. The USAF sets internal goals for D2D/M2D ratios by weapon system and support area. For instance, a stated goal D2D/M2D ratio for active duty mobility forces is 1:2 and for reserve component mobility forces is 1:5.[1] Per OSD policy, a unit is considered deployed when the majority of the unit is engaged in a Secretary of Defense–approved operational requirement away from home station. A reserve component unit is considered mobilized per the terms of Secretary of Defense–approved orders so stating.

RAND obtained D2D/M2D data from Air Mobility Command's (AMC's) A3 and A9 directorates and also used data from the Global Decision Support System and the Logistics Installations and Mission Support–Enterprise View to support the analysis that follows.

---

[1] Action officers and stakeholders participating in study interim progress reviews routinely mentioned this.

Deployment-to-Dwell/Mobilization-to-Dwell Management

Although the USAF as a whole adheres to the general principles of the D2D/M2D framework, specific D2D/M2D management varies by community. Different parts of the USAF calculate D2D/M2D ratios differently depending on how specific forces are managed and operated. For example, Air Combat Command and the combat air forces typically calculate a "historic" D2D/M2D ratio, which uses the duration of the second-to-last deployment and the dwell time between that deployment and the most recent deployment. AMC and the Mobility Air Forces (MAF) often report D2D and M2D using a "today" approach for calculating the D2D/M2D ratios using the number of deployed crews divided by the total number of crews, by weapon system and component.

In other parts of the USAF, the line between a deployment and a dwell period is blurry. The service members in the remotely piloted aircraft, intercontinental ballistic missile, intelligence, space, and cyber communities all "deploy in garrison." In other words, they perform operational missions from home station. The high demand for the services of these communities by the combatant commands, coupled with the difficulty in measuring OPTEMPO using traditional D2D/M2D-type metrics, could be affecting readiness.

It is also challenging to apply M2D policy to the Air National Guard and Air Force Reserve because a significant portion of reserve component deployments in the USAF are voluntary, yet M2D policy applies only to mobilizations. The USAF plans for about 25–30 percent of forces it needs to deploy to come from reserve component volunteers. Yet if they do volunteer for a deployment, it does not count under current policy, and they are subject to being scheduled to deploy as if they had not deployed voluntarily "out of cycle." Traditionally, the volunteerism rates vary from mission area to mission area and are affected by the state of the economy. Active duty forces then make up the difference between what is required and what remains.

The USAF rarely pushes back on emerging requirements because of D2D/M2D constraints, except in a few key specialties or weapon systems. The prospective impacts on readiness resulting from the stress of meeting emerging requirements given D2D/M2D constraints can be challenging to assess because of significant differences among weapon systems and specialties. In some communities, deployments can generate readiness because they act as a "forcing function" to get units to complete all of their readiness requirements before departure. In other communities, deployments consume readiness because they limit the ability to train to certain mission sets.

Moving Toward a New Force Generation Model

Senior leaders within the USAF have acknowledged the challenges posed by managing force generation differently across the USAF, particularly as it aligns with the DoD strategy to better prepare for great-power conflict.[2] As of this writing, the USAF is transforming how it presents forces, shifting from the Air Expeditionary Force model, which permits deployments of sub-squadrons and even piecemeal deployments of individuals, to a new force presentation model built around deploying cohesive squadrons as a whole. This new force presentation model, called the *Air Force Force Generation construct*, will be cyclical, rotating units through different stages of readiness and availability.[3] Several communities within the USAF, including KC-135 and C-130J forces, have already shifted to this new way of doing business. In an interview supporting this research, AMC officials told us that C-17 forces and KC-46 forces will soon follow suit.[4] Although the intent is to align the force presentation model with the D2D/M2D framework, it remains to be seen how the current transformation of the way the USAF presents forces will affect the management of D2D/M2D policy.

Deployment to Dwell/Mobilization to Dwell in the Mobility Air Forces—KC-135 Case Study

The basic fighting unit in the USAF is the squadron, and, in many instances, the USAF manages forces at the squadron level. AMC employs some of its forces at the squadron level, although typically just for operational deployments. Official deployments are those managed through the GFMAP process. However, MAF units frequently support day-to-day activities outside the GFMAP process via taskings from USTRANSCOM because of the long-established use of mobility forces to support geographic combatant command (GCC) requirements via USTRANSCOM's tasking authority over those forces. If USTRANSCOM did not fill the requirements, the demand would be tasked through the usual GFMAP processes.

Most of these taskings do not require a full squadron; therefore, AMC tracks utilization of crews rather than squadrons. For example, C-5 units are never deployed or tasked as squadrons, so a squadron-based metric would be problematic. Other MAF units are sometimes deployed or tasked as squadrons or partial squadrons. Therefore, AMC reports D2D and M2D using a crew-based ratio of deployed crews divided by the total number of crews, by weapon system and component.

The KC-135 Stratotanker is the USAF's primary aerial refueling asset and is vitally important to today's warfighter. It has played a crucial role in operations spanning as far back as the Vietnam War. More recently, it has provided critical aerial refueling capabilities in support of Operations Iraqi Freedom and Enduring Freedom. The KC-135's importance within the

[2] Statements by Air Force Chief of Staff Gen Charles Brown, as reported in Brian W. Everstine, "CSAF Plans a Better Deployment Model," *Air Force Magazine*, October 1, 2020.

[3] Department of the Air Force, "AFFORGEN Update: Sustainable Force Offerings," briefing, January 25, 2021.

[4] Telephone interview, AMC A3O officials, October 2020.

MAF and its ongoing use in support of operations make it a good case study for examining D2D and M2D implications for the MAF.

Figure 4.1 shows the breakdown of KC-135 squadrons, primary aircraft authorized (PAA), and crews by component. Approximately two-thirds of the KC-135 force structure is in the reserve component. The active duty squadrons each have 12 to 15 PAA, while the reserve component squadrons typically have eight to 12 PAA.

Several observations about the implications of a D2D/M2D policy for the KC-135 platform emerge upon examination of just the demographics of the KC-135 force structure:

- Because the majority of the KC-135 fleet is in the reserve component, accessing the majority of USAF KC-135 capability for deployment requires a combination of volunteerism and mobilization.
- Because the reserve component threshold M2D ratio is higher than the D2D threshold, less capability is available for the same force structure (the KC-135 platform) compared with weapon systems with a larger active component proportion.
- Active component squadrons are larger than reserve component squadrons, and these differences must be taken into account when these units are being scheduled.
- Management of D2D and M2D can be a complex mix of active and reserve component crews and shared aircraft in six associate units.

Although examining basic KC-135 demographics provides some insights into the implications of a D2D/M2D policy, additional analysis is required to better understand how changes to the policy, or the introduction of a different policy, might affect the ability of the KC-135

FIGURE 4.1
KC-135 Squadrons, Primary Aircraft Authorized, and Crews, by Component

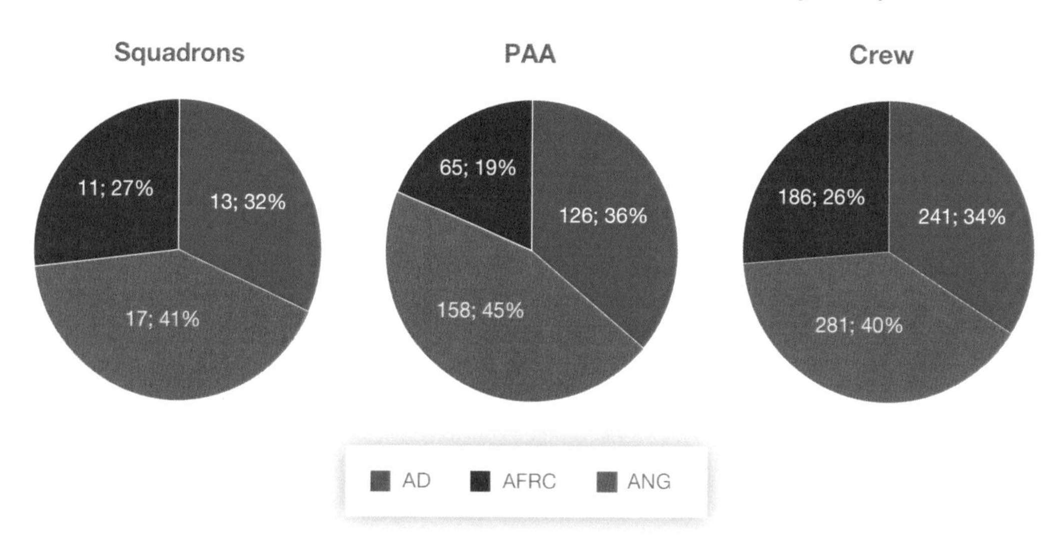

NOTE: AD = active duty; AFRC = Air Force Reserve Command; ANG = Air National Guard.

fleet to satisfy mission demands. In the following sections, we introduce a methodology and illustrative analysis to highlight some of these considerations.

Methodology

RAND developed the MAF Force Generation (MAFORGEN) Scheduling Model to examine how D2D/M2D policies affect the ability of the KC-135 fleet to meet different demand scenarios. The model uses Microsoft Excel for inputs and outputs and Visual Basic for Applications code for processing. As shown in Figure 4.2, the model assigns available squadrons ("supply") to deployment requirements ("demand") subject to a set of policy constraints and provides the resulting deployment schedule, including any unmet requirements, as the output.

Inputs

Demand is specified as a series of deployment requirements. Each deployment is characterized by the mission design series type required, a start day, a deployment length, the number of aircraft required, and the number of crews required. Supply is specified as the set of squadrons available for deployment. Each squadron is characterized by the mission design series, component, number of aircraft, crew ratio, and current D2D and M2D. The primary policy options used in this analysis were the D2D/M2D limits specified by component (active duty forces, the Air National Guard, and the Air Force Reserve). The model also includes the ability to incorporate a rotational readiness system in which squadron deployments are limited by rotational constraints; this option was not used in our analysis but could be included in future studies.

Model Logic

The model seeks to fill all deployments from the available supply of squadrons subject to the D2D/M2D policies. It allocates supply starting with the first deployment and moving down

FIGURE 4.2
RAND Mobility Air Forces Force Generation Scheduling Model

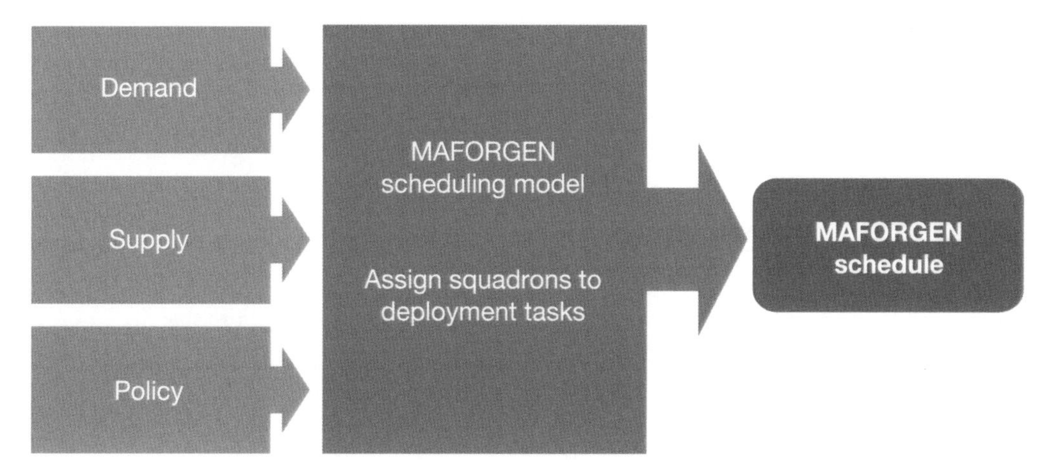

the deployment list in sequence until all deployments have been evaluated and matched if possible. For each deployment, the model first identifies a list of all units that could potentially fill the deployment, removing all from consideration that do not meet the requirements either because of a policy setting or because the attributes of the unit do not match the required attributes of the deployment. Of the units that remain viable for a given deployment, the model assigns the squadron with the highest current D2D and M2D ratios (meaning the squadron that has been home the longest in comparison to its last deployment).

Output

The primary model output is the deployment schedule, which shows the number of deployments satisfied, including specific squadron assignments, and the number of unfulfilled deployments.

Analysis

We framed the analysis of D2D/M2D policy ratios using MAFORGEN to predict filled versus unfilled deployed demand for KC-135 squadrons using unconstrained, current threshold, and current goal ratios. We used threshold ratios of 1:2 for the active component and 1:4 for the reserve component and goal ratios of 1:3 for the active component and 1:5 for the reserve component. We then ran MAFORGEN using the current KC-135 force structure against a hypothetical constant demand for ten KC-135 squadrons across a three-year span.[5] For the purposes of this analysis, we assumed that each deployment required seven aircraft and 13 crews and lasted four months. We also assumed that a squadron could be tasked to fill only one deployment at a time.[6] This resulted in a total demand of 90 squadron deployments in 36 months. In each excursion, we assumed no competing requirement for the force to meet USTRANSCOM demands outside of deployments and did not consider any at-home readiness training requirements.

Hypothetical Steady-State Ten-Squadron Demand

In the baseline case, not surprisingly, given a 41-squadron,[7] 349-aircraft fleet size, an unconstrained USAF KC-135 force structure was able to meet 100 percent of a constant ten-squadron demand signal, as shown in Figure 4.3.

However, the MAFORGEN-predicted results changed when a policy constraint requiring the force to not exceed current threshold D2D/M2D ratios (1:2, 1:4) was imposed. Like the

---

[5] We created hypothetical tanker demands to keep the report unclassified and to provide analytical results that illustrate differences among various policy options.

[6] We considered the possibility of analyzing the creation of temporary new units using the left-behind parts of two or more deployed squadrons. This is commonly referred to as *rainbowing* units to fill a tasking. Although this has been done in the past, it is not preferred on the basis of unit integrity, and AMC told us that it does not intend to make use of rainbowing in the future.

[7] Thirty-five of these squadrons are traditional squadrons, and six are associate squadrons.

first case, in this case we did not consider competing USTRANSCOM or readiness demands on the fleet. Constrained by the threshold ratios, the force was able to meet 74 of 90 total deployments (82 percent), as Figure 4.4 illustrates.

FIGURE 4.3

Deployments Filled with Unconstrained Deployment-to-Dwell/Mobilization-to-Dwell Policy Ratio

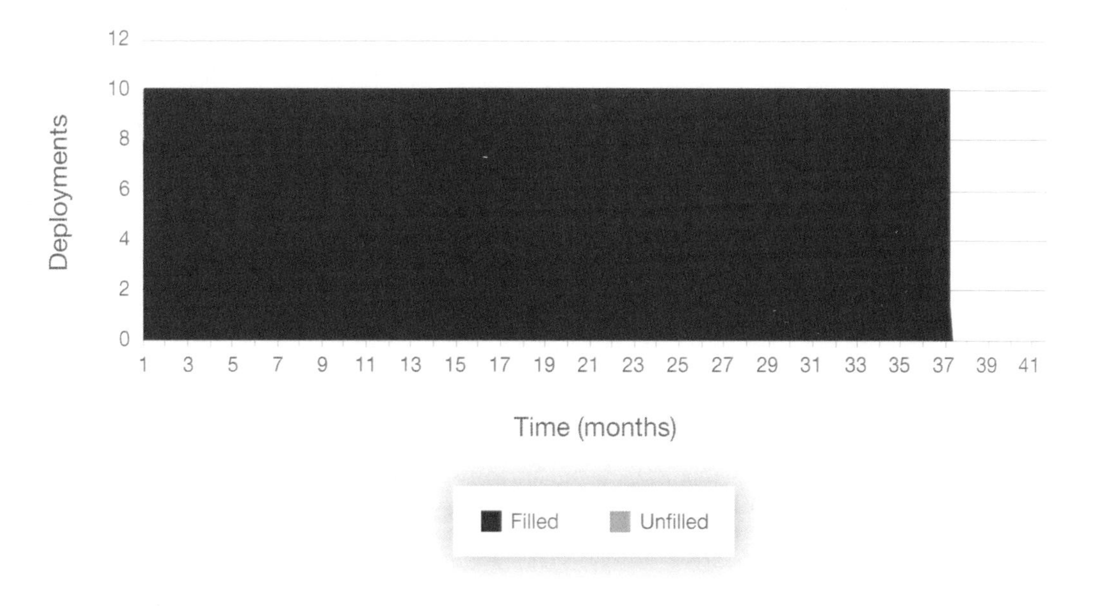

FIGURE 4.4

Deployments Filled with Threshold Ratios
(1:2 for the active component, 1:4 for the reserve component)

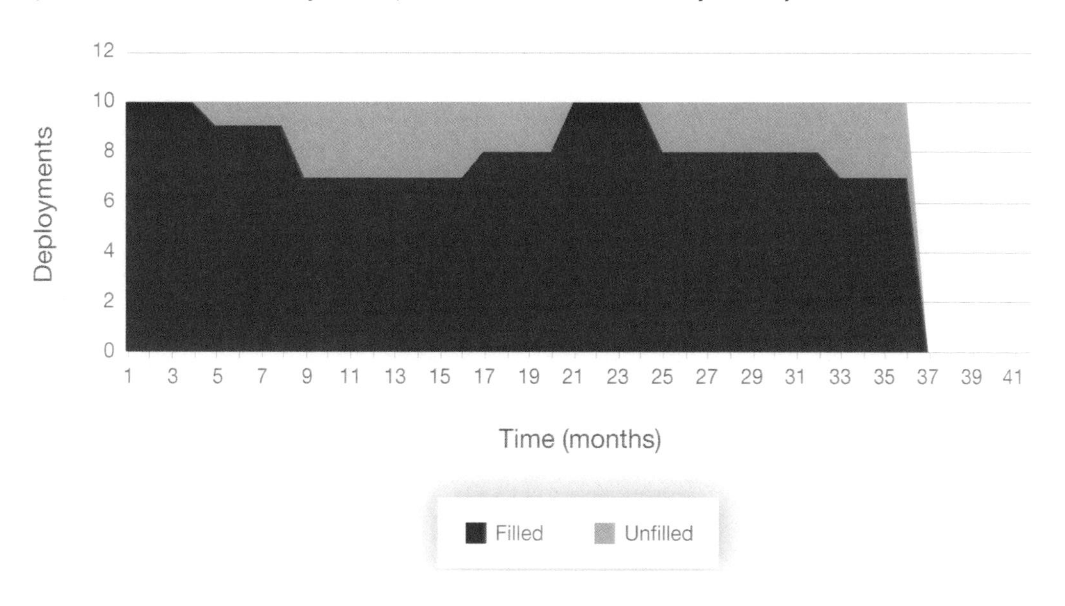

Lastly, we imposed the current D2D/M2D goal ratios of 1:3 for the active component and 1:5 for the reserve component against the same ten-squadron demand without consideration of competing USTRANSCOM or readiness demands. Figure 4.5 shows that the force was able to meet 67 percent of the demand under these conditions.

As expected, as the required dwell in the D2D/M2D policy ratios increases, fewer deployment demands can be met, thus reducing the total availability of units to deploy.

Excursion 1: Hypothetical Major War and Transition to a Steady State

For the first excursion, we analyzed the effect of meeting the same constant ten-squadron demand, using the goal ratios of 1:3 and 1:5, but this time after a hypothetical major war in which all active component and half of reserve component squadrons returned from the war on the first day of the same 36-month steady-state period analyzed in the baseline cases. With fewer squadrons available at the start of the excursion, the fill rate dropped to 53 percent, reflecting the effects of a major war on availability and the scheduling turbulence that results from transitioning from a major war effort to a steady state (Figure 4.6).

Excursion 2: Hypothetical Demand Surge

Next, we analyzed a scenario that illustrates the idea of building a strategic reserve. In this case, we examined what would happen if the force were to experience a 28-month period of steady-state demands of six squadrons, three squadrons, or one squadron followed by a major war requiring a surge to 35 squadrons while still honoring the goal ratios of 1:3 and 1:5.

FIGURE 4.5
Deployments Filled with Goal Ratios (1:3 for the active component, 1:5 for the reserve component)

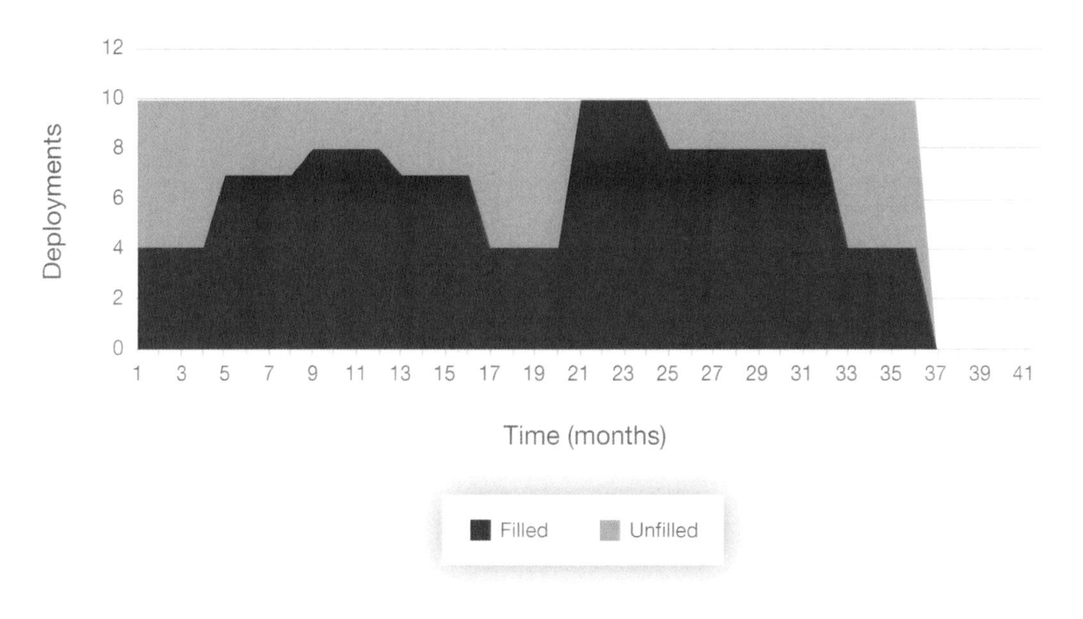

FIGURE 4.6
Deployments Filled After Transition from Major War to Steady State with Goal Ratios
(1:3 for the active component, 1:5 for the reserve component)

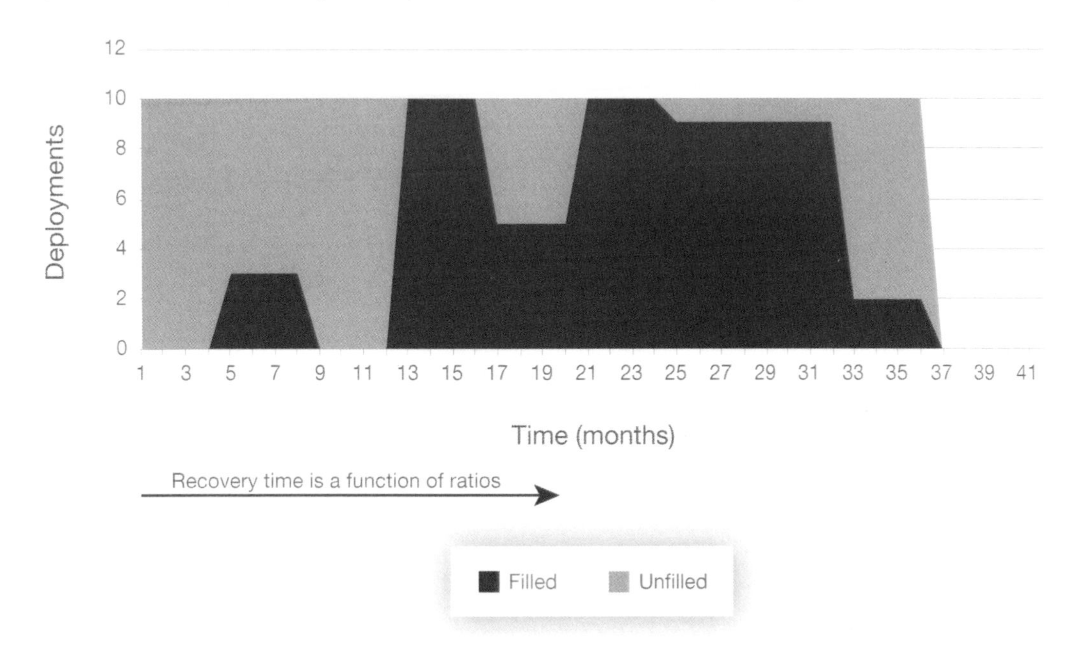

Under these assumptions, the force could fill only 13 (37 percent) of the 35 units required to support the surge after sustaining a 28-month span requiring six squadrons (Figure 4.7). We recognize that, in a real major war, the USAF would send as many units as was required, regardless of the status of any particular unit's dwell vis-à-vis D2D/M2D policy. Thus, in a sense, the red area of unfilled demand in the figure illustrates how many units would require policy waivers and the number of units that would have to be sent out again after not fully recovering from their last deployments.

Next, we supposed a constant three-squadron demand under the same conditions as described for the six-squadron case. Figure 4.8 shows that the force was able to meet 24 (69 percent) of the 35 squadron demands in a surge for a major war.

Last, we ran MAFORGEN using the same conditions but with a prewar constant demand of only one squadron. Figure 4.9 shows that the force was able to meet 31 (89 percent) of the 35 required squadrons in the surge.

In general, as the prewar demand goes down, the ability to surge to meet a crisis increases. We think that these three cases illustrate the concept of a strategic reserve. Lessening the demand on the force for a stretch of time builds capacity to surge in the event of a crisis without needing to make use of units that might not be fully ready to go.

FIGURE 4.7

Deployments Filled in a 35 Demand Surge Following a Six-Squadron Steady-State Demand with Goal Ratios (1:3 for the active component, 1:5 for the reserve component)

FIGURE 4.8

Deployments Filled in a 35-Demand Surge Following a Three-Squadron Steady-State Demand with Goal Ratios (1:3 for the active component, 1:5 for the reserve component)

FIGURE 4.9
Deployments Filled in a 35-Demand Surge Following a One-Squadron Steady-State Demand with Goal Ratios
(1:3 for the active component, 1:5 for the reserve component)

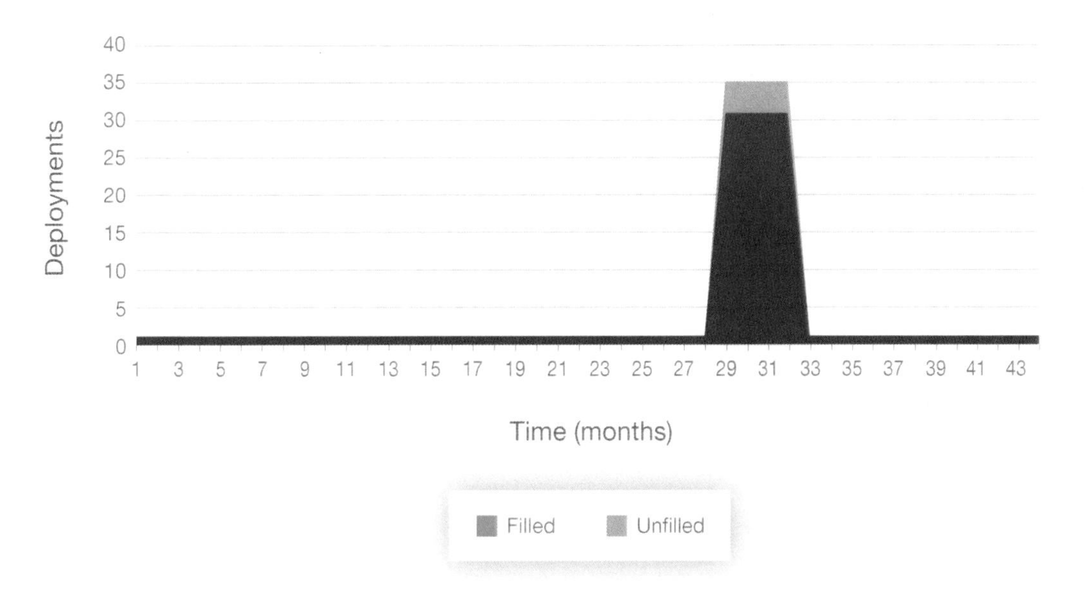

Exploring Other Metrics—Task to Dwell

Up to this point, we have not considered competing demands for KC-135 capability that come from USTRANSCOM or for USAF-required training to maintain readiness. USTRANSCOM argues that the MAF would be best managed using a metric that it calls *task to dwell* (T2D). USTRANSCOM uses T2D to manage its day-to-day operations by balancing the competing risk to force (readiness for future needs) and risk to mission (current needs). Mathematically, T2D is different from D2D and M2D. USTRANSCOM defines T2D as the number of tasked crews per day divided by the number of nontasked crews, averaged over the past 90 days.[8] T2D measures the daily ratio of the force that is "busy" conducting missions for other GCCs versus the force that is at home and able to either train or attend to other dwell-time matters.

In this metric, a tasked crew includes all crews tasked within the GFMAP and Secretary of Defense Orders Book processes and all of those tasked by USTRANSCOM outside these processes but which would be tasked by the Secretary of Defense if USTRANSCOM were unable to fill the mission. At issue is how to balance taskings with service readiness requirements and at-home needs. Figure 4.10 shows the comparison between KC-135 D2D/M2D ratios and their corresponding T2D ratios from 2009 to 2016. The data are a compilation

[8] USTRANSCOM, J3, "Policy Proposal for the Addition of Task-to-Dwell (T2D) as Tempo and Readiness Metric for USTRANSCOM-Assigned Mobility Air Forces (MAF)," memorandum to the Under Secretary of Defense for Personnel and Readiness, July 30, 2020.

FIGURE 4.10
RAND-Computed KC-135 Fleet Deployment-to-Dwell/Mobilization-to-Dwell Ratios Versus Task-to-Dwell Ratios

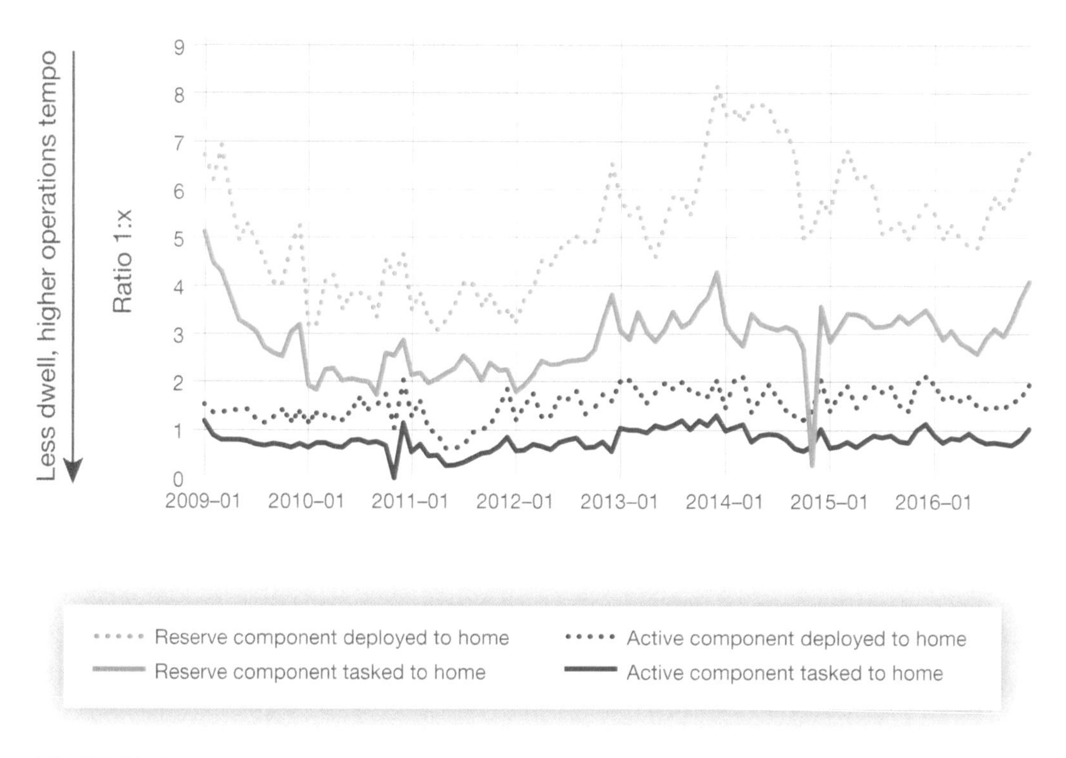

SOURCE: RAND analysis of 2009–2016 data from AMC's Global Decision Support System and the Logistics Installations and Mission Support–Enterprise View.
NOTE: Calculated based on tails deployed versus tails home using a 1.5:1 crew ratio for deployments.

from AMC's Global Decision Support System and the Logistics Installations and Mission Support–Enterprise View. Because neither of these data sets contains crew information, we used aircraft tails as a proxy, using 1.5 crews per aircraft for deployed tails and 1 crew per aircraft for nondeployment, but operational, tails.[9]

As shown, T2D ratios were numerically lower than D2D/M2D ratios at all times, and the KC-135 fleet was quite busy during these years. Figure 4.11 shows more-recent T2D ratios, as computed by AMC's A9 Analysis directorate. The gray lines across the chart show 1:1, 1:2, 1:3, and 1:4 T2D ratios. Recent KC-135 fleets were less busy than those from 2009 to 2016, and T2D ratios from these years generally fell within AMC's target ratio of between 1:2 and 1:3.

For analysis purposes, we used MAFORGEN to compare D2D/M2D and T2D policies by assessing predicted fill rates against the same ten-squadron demand signal used earlier.

---

[9] Whereas garrison units have higher crew ratios (approximately 2 to 2.5 crews per plane), it is typical for the crew ratio for a deployment to be 1.5 crews per plane and for a nondeployment operational mission to be 1 crew per plane.

FIGURE 4.11
Air Mobility Command A9–Produced KC-135 Task to Dwell, January 2019 to October 2020

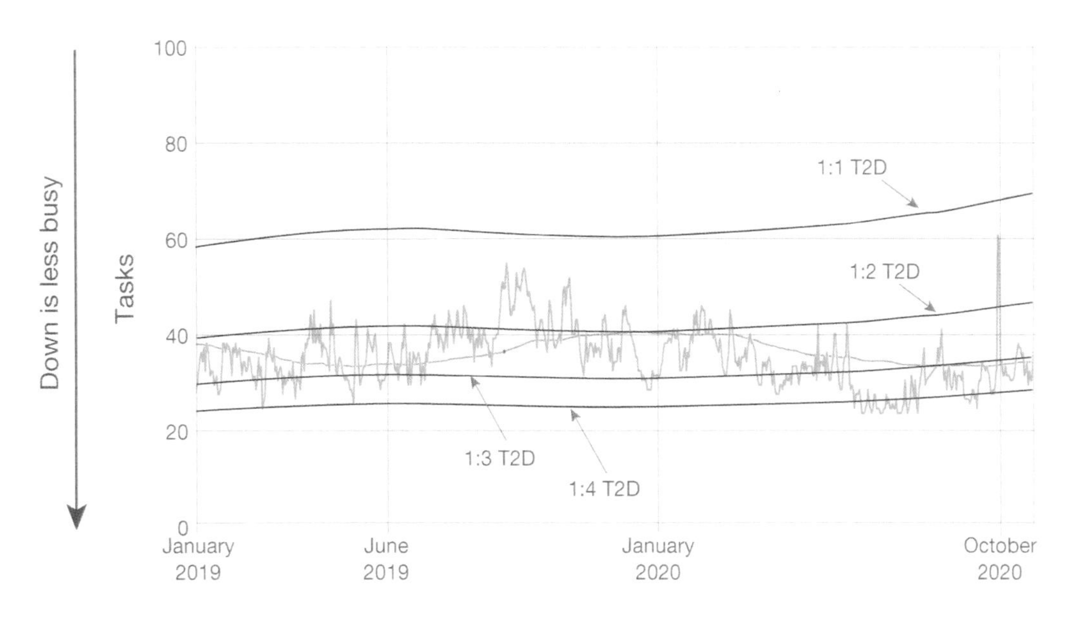

SOURCE: AMC/A9.
NOTE: The green line indicates daily tasks, and the red line indicates the 180-day T2D.

Recall that, when we did not model any USTRANSCOM demand and used threshold D2D/M2D ratios of 1:2 for the active component and 1:4 for the reserve component, the force was able to meet 74 (82 percent) of 90 deployments (see Figure 4.4). In this excursion, we added a historically representative USTRANSCOM demand signal requiring 10 percent of the KC-135 force to support other demands apart from Secretary of Defense–approved deployments to the ten-squadron deployment demand. We also changed the policy from D2D/M2D ratios to the T2D policy ratio of 1:2 for the active component and 1:4 for the reserve component. With these conditions, the force could meet only 49 of 90 deployments (54 percent of the deployment demand), as shown in Figure 4.12. Not surprisingly, the force was less able to fill deployment demand while simultaneously meeting other demands and enforcing a capacity-limiting policy ratio.

Figure 4.13 shows the relationship between T2D and D2D/M2D when USTRANSCOM tasks 10 percent of the fleet to do "T2D-counting" missions outside of deployments. T2D shows a numerically lower (meaning busier) ratio than D2D and M2D in all cases. What it shows, for instance, is that at the goal D2D/M2D ratio of 1:5 for reserve forces, those mobility forces would experience a T2D ratio of about 1:2.75 if 10 percent of those forces were also conducting missions for USTRANSCOM.

This is USTRANSCOM's point in proposing T2D as an official metric for use under a D2D/M2D policy. USTRANSCOM implicitly argues that the purpose of the D2D/M2D policy is to limit OPTEMPO and time away from home for service members in the MAF.

FIGURE 4.12
Deployments Filled with Task-to-Dwell Ratios of 1:2/1:4 Plus 10-Percent U.S. Transportation Command Taskings

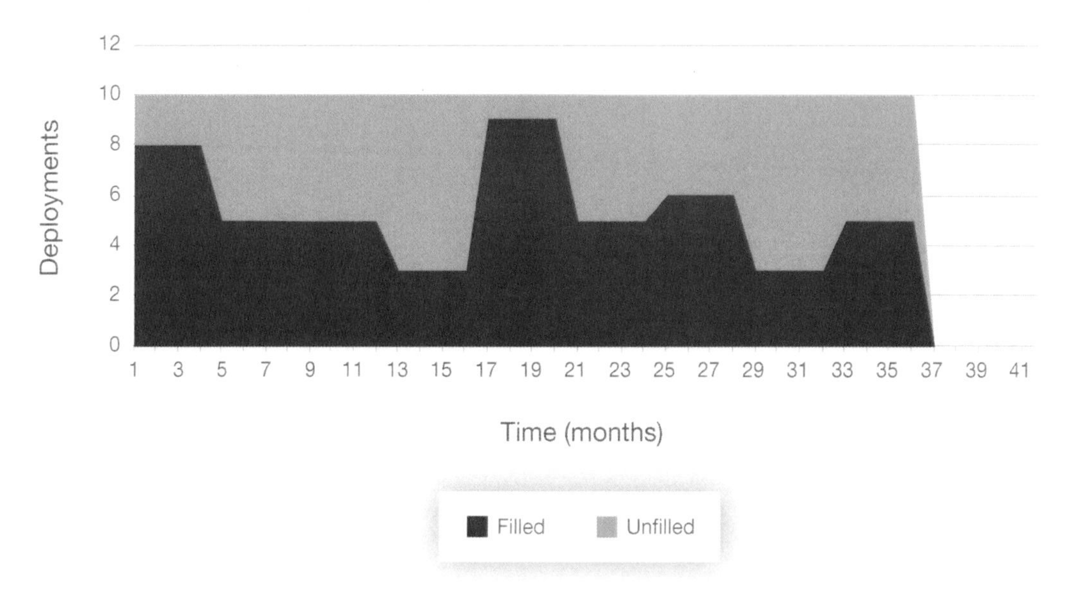

FIGURE 4.13
Relationship of Deployment to Dwell to Task to Dwell Assuming 10 Percent of Fleet Tasked by U.S. Transportation Command to Conduct Missions Outside of Deployments

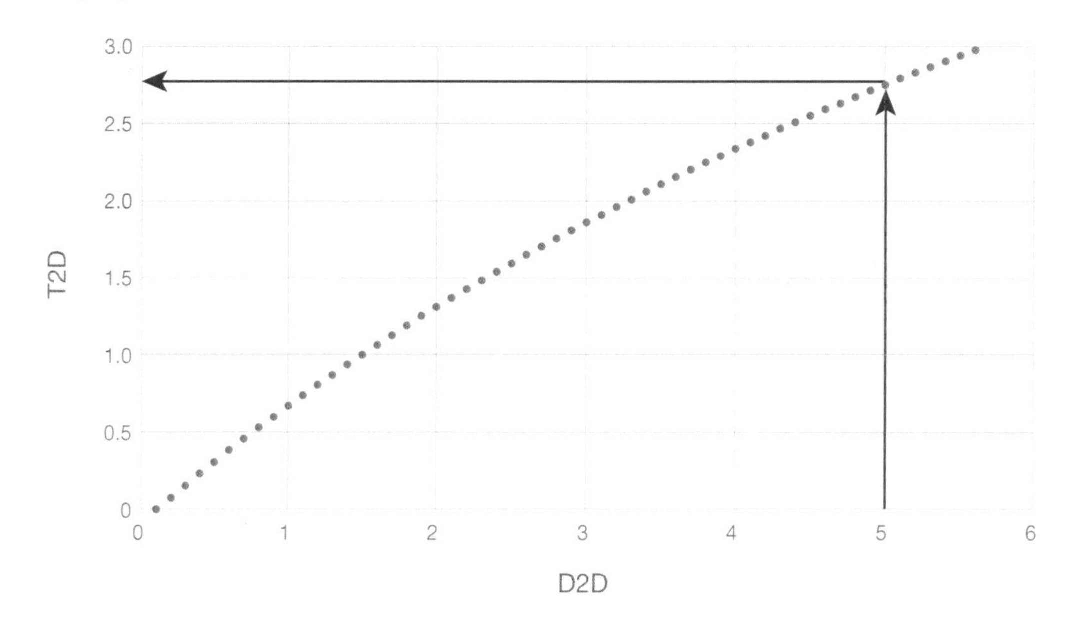

However, D2D and M2D do not capture how busy the MAF really are, and the concern is that if the MAF were to be tasked up to the stated D2D/M2D ratio limits, the resulting actual T2D ratio might overstress the force. These results highlight a key policy question: What kind of missions should count against a D2D/M2D metric?

Summary

We offer the following observations, which are based on our examination of how the D2D/M2D policy is implemented for the MAF and our analysis of the KC-135 fleet:

- The USAF's KC-135 force structure is adequate to meet current Secretary of Defense deployment demands using a threshold 1:2, 1:4 policy. However, its ability to meet those demands decreases as policy ratios increase numerically or if more mission types are included in what "counts" while the same ratios are enforced.
- The MAF are not strictly implementing the D2D/M2D policy as written because they use T2D as their primary metric for the daily management of OPTEMPO. They assert that a T2D metric better meets the intent of the D2D policy because of the number of missions that the force must meet that are not captured by the D2D policy.
- Alternative metrics (like T2D) might be better for protecting dwell time and readiness, but
  - goal and threshold ratios need to be set carefully for a clearly stated purpose and might need to vary by platform
  - what counts as deployed or tasked must be adjudicated—especially activities not approved by the Secretary of Defense.

CHAPTER FIVE

Deployment- and Mobilization-to-Dwell Policy in the U.S. Navy

In this chapter, we describe how the U.S. Navy implements the D2D/M2D policy and explore possible limitations, constraints, and opportunities that this policy can have for the Navy using the aircraft carrier (specifically, the ship's company) as a case study. The chapter begins with a description of how the policy manifests in the Navy, and special attention is paid to Navy deployment and maintenance cycles. This discussion is followed by an analysis of recent historical deployment and maintenance cycles to assess observed D2D for the aircraft carrier.

Policy

Unlike the Marine Corps, the Navy does not have a separate D2D/M2D policy. However, the Navy does have policies that enable control over the pace of a Navy unit's operations. The first relevant policy is Office of the Chief of Naval Operations (OPNAV) Instruction (OPNAVINST) 3000.13E, *Navy Personnel Tempo and Operating Tempo Program*.[1] This instruction provides the policy, procedures, and reporting requirements regarding the PERSTEMPO of individual sailors and the OPTEMPO of deployable units, and it revised the limits set by OPNAVINST 3000.13D. Although D2D for individuals is not specifically tracked in the Navy, the PERSTEMPO program tracks deployment and nondeployment events for individual sailors, and this tracking, combined with the Optimized Fleet Response Plan (OFRP), directly contributes to the Navy's existing D2D/M2D ratios.

Navy Personnel Tempo and Operating Tempo Program

OPNAVINST 3000.13E states that the Navy must monitor and evaluate how deployment lengths and schedules affect sailors' quality of life and emphasizes the need for accurate data. As defined in the policy, PERSTEMPO events fall into either deployment or nonde-

[1] OPNAVINST 3000.13E, *Navy Personnel Tempo and Operating Tempo Program*, Washington, D.C.: Department of the Navy, January 27, 2021.

ployment categories, and the PERSTEMPO system ensures that each sailor has a continuous PERSTEMPO deployment counter.[2]

The instruction also limits operational deployment length for a Navy unit to 220 days and sets high-deployment waiver thresholds. The one-year high-deployment waiver threshold is a maximum of 220 days deployed in a 365-day period, and the two-year high-deployment waiver threshold is a maximum of 400 days deployed in a 730-day period. If PERSTEMPO limits will be exceeded, advance approval is required. Per 10 U.S.C. 991, the Secretary of Defense must approve any exceptions to PERSTEMPO limits, but this approval was suspended with the 2001 national security waiver, which allows the first flag or general officer, senior executive, or Senate-confirmed appointee in the chain of command to approve exceeding PERSTEMPO limits. The instruction also includes the statement that involuntary active duty for reserve members must not exceed 12 months.

OPNAVINST 3000.13E defines "deployment" PERSTEMPO events as operations, exercises, unit training, mission support and temporary duty, and training at home station or in the "local operating area of a ship or vessel."[3] "Non-deployment" PERSTEMPO events are individual training, duty in home station or home port, hospitalization in an area of permanent duty station or home port, disciplinary events, inactive duty training for reserve component members, muster duty for reserve component members, and funeral honors duty.[4]

All Navy activities must report all PERSTEMPO events in the PERSTEMPO system, and the policy requires leaders to be "personally involved" with this reporting process. Navy Operational Support Center commanders must report reserve component PERSTEMPO events.[5]

OPNAVINST 3000.13E also explains OPTEMPO policy and reporting requirements, noting that a unit is "either in dwell or on an operational deployment."[6] Per the policy, the maximum unit operational deployment length is 220 days, the same as the PERSTEMPO limit. The Chief of Naval Operations must approve any deployment length exceeding 220 days for a unit, and the Secretary of Defense must approve any length over 365 days for a unit.[7]

The policy further defines *dwell*, noting that the dwell limit is 1:2, with a goal of 1:3, for the active component and 1:4, with a goal of 1:5, for the reserve component. Per OPNAVINST 3000.13E, the 2019 contingency planning guidance, and the FY 2019–2020

---

[2] OPNAVINST 3000.13E, 2021, Enclosure 1, pp. 1–2.

[3] OPNAVINST 3000.13E, 2021, Enclosure 1, p. 3.

[4] OPNAVINST 3000.13E, 2021, Enclosure 1, pp. 3–4.

[5] OPNAVINST 3000.13E, 2021, Enclosure 1, p. 5.

[6] OPNAVINST 3000.13E, 2021, Enclosure 2, p. 1.

[7] OPNAVINST 3000.13E, 2021, Enclosure 2, p. 1.

Global Force Management Implementation Guidance, exceeding the D2D or M2D ratio requires the approval of the Secretary of Defense.[8]

The policy outlines exceptions to this rule:

- Navy multicrewed active component forces built for continuous presence, such as Blue and Gold crews on nuclear-powered ballistic missile submarines, are not allowed to exceed 1:1 D2D ratios without approval from the Secretary of Defense.
- Navy units designed specifically for a 1:2 D2D ratio, limited to that ratio by force design, are not considered to be operationally deployed until they assume mission or command, and the D2D count stops when command or mission is relinquished, exclusive of transit time or turnover (although these still register as PERSTEMPO events). This specification ensures that units locked into a 1:2 D2D design do not trigger Secretary of Defense approval processes unless their schedules also exceed the designed D2D.[9]

OPNAVINST 3000.13E also requires the Navy to track cumulative underway days for sailors and to report for a three-year period (two years back and one year forward), counting every day that a unit is underway or away from its permanent station as a unit. Cumulative underway days must not exceed 547 days in a 1,095-day period except for submarine units, which have a ceiling of 613 days because of maintenance cycle demands. This accrual starts on the first day underway and continues until return to homeport, and exceeding the limit on cumulative underway days requires Navy component commander approval and Chief of Naval Operations notification.[10] For units with rotational crews (such as nuclear-powered ballistic missile submarines), D2D is tracked and reported for individual crew members; units that deploy as detachments track and report D2D for the detachment.[11]

OPNAVINST 3000.13E requires all Navy commands, activities, and units to ensure that their next supervisor is provided PERSTEMPO and OPTEMPO data to enable compliance. Forward-deployed naval forces are subject to the same limits, but D2D ratios include only operational deployment time in support of commanders outside these forces' area of responsibility.

Optimized Fleet Response Plan

The Navy's PERSTEMPO policy and D2D/M2D ratios as presented in OPNAVINST 3000.13E align with OFRP. OFRP is an operational framework that is designed to optimize the return on training and maintenance investments, maintain sailor quality of service, and ensure that units and forces are certified in defined, progressive levels of employable and deploy-

[8] OPNAVINST 3000.13E, 2021, Enclosure 1, p. 2.

[9] OPNAVINST 3000.13E, 2021, Enclosure 2, pp. 1–2.

[10] OPNAVINST 3000.13E, 2021, Enclosure 2, p. 2.

[11] OPNAVINST 3000.13E, 2021, Enclosure 2, p. 3.

able capability. OFRP contains four phases (maintenance, basic, integrated or advanced, and sustainment), and an OFRP cycle begins at the start of each maintenance phase and ends at the start of the next maintenance phase. Cycle lengths are designed to meet OFRP goals for each force element.[12]

The maintenance phase, the start of the OFRP cycle, applies to all deployable Navy forces, and this phase is where most major shipyard or depot-level repairs, upgrades, and modernization efforts occur. Inspections, certifications, visits, and various training efforts continue through this phase; the OFRP policy notes that, for a carrier, the maintenance phase can take up to 16 months. The basic phase includes core capabilities and skills, with inspections, certifications, assessments, and training. Personnel, equipment, and various supply efforts are completed during this phase.[13]

The integrated phase, which also includes required training, inspections, and certifications, is designed to integrate units and staffs into strike groups to achieve readiness and prepare crews for coming deployments.[14] This phase can include specialized training as needed and ends with certification for deployment in a variety of environments. The advanced phase is primarily for those Navy units that do not belong to a deploying group already and allows for additional inspections, certifications, training, and more to build deployment-ready forces.[15]

Finally, the sustainment phase starts at the conclusion of the integrated or advanced phase and ends with the start of the next maintenance phase—and it includes deployments. Per the OFRP policy, a variety of factors (unit, command requirements, anticipated tasking, funding, and readiness level) can inform the length and substance of the sustainment phase.[16]

OFRP's scheduled maintenance and training periods create built-in opportunities for dwell, making D2D and the force generation cycle concomitant. As one Navy official stated during a discussion for this project, "The ship is not going to be in dwell and unavailable because of the D2D of the people on the ship." Maintenance schedules tend to face delays, and this reality enhances the Navy's ability to meet D2D goals, but these delays can result in unpredictable OPTEMPO. The Navy's Fleet Forces Command is responsible for executing OFRP.

---

[12] OPNAVINST 3000.15A, *Optimized Fleet Response Plan*, Washington, D.C.: Department of the Navy, November 10, 2014, pp. 1–2.

[13] OPNAVINST 3000.15A, 2014, p. 4.

[14] It is noteworthy that the Navy achieves readiness outside of the integration phase prior to deployment. Navy leadership has indicated that deployments increase readiness; see "Operations in South China Sea Increase Readiness of Squadrons: US Navy," *Business Standard*, last updated July 6, 2020.

[15] OPNAVINST 3000.15A, 2014, p. 5.

[16] OPNAVINST 3000.15A, 2014, p. 5.

U.S. Navy Reserve Policy

Unlike the tracking of unit D2D for the active component, the Navy's M2D policy is largely applied to individuals. After 9/11, the Navy began using IAs to source the joint force and to help address the challenges presented by a high tempo during operations in Iraq and Afghanistan. Although IAs were initially sourced from both the active component and the reserve component, this shifted over time to fall heavily onto the reserve component. According to testimony from Vice Admiral Mustin, Chief of the U.S. Navy Reserve (USNR), over the past few years, USNR has sourced over 75 percent of all IA requirements for the Navy. Most of these, according to his testimony, have fallen outside the maritime domain.[17]

USNR leadership is striving to improve reserve readiness by cutting back on support to nonmaritime missions. Vice Admiral Mustin testified in May 2021 that, under the current DoD policy, USNR has "9% of the force mobilized on IA missions and 18% sequestered in dwell, [so] 27% of the Force is effectively fenced from surge mobilization as a result of IA demand."[18] As a result, Vice Admiral Mustin is endeavoring to realign USNR employment through his "mobilize the force" line of effort, articulated in his "Navy Reserve Fighting Instructions 2020: Design/Train/Mobilize" guidance.[19] After a one-year deployment, approval from the Secretary of Defense must be obtained for any reserve component member to deploy with their unit if it has been less than four years since their last mobilization; this policy has also significantly affected the depth and readiness of reserve component units.

While the USNR is pushing to move away from the IA, nonmaritime mobilization model of the past two decades, reserve unit activations generally remain limited to Navy Special Warfare units, aviation (such as Strike Fighter squadrons, Electronic Attack squadrons, Fleet Logistics Support squadrons, Patrol squadrons, and Fighter Composite units), Navy Mobile Construction battalions, Expeditionary Logistics, and Coastal Riverine squadrons. Beyond these exceptions, even if a unit is tasked, unit members are typically mobilized as individuals. Additionally, relevant to the JFE explored in this chapter, USNR units do not routinely embark on carriers, nor are positions programmed to carriers or associated air wings. IA manpower is also not currently directed toward carrier strike groups.

The Navy Aircraft Carrier

The Navy JFE selected for this analysis, the aircraft carrier (CVN), is the "centerpiece of America's Naval forces."[20] Aircraft carriers remain go-to capabilities for policymakers

[17] John B. Mustin, Chief of the U.S. Navy Reserve, "Fiscal Year 2022 National Guard and Reserve," statement before the Subcommittee on Defense, Committee on Appropriations, U.S. Senate, Washington, D.C., May 18, 2021.

[18] Mustin, 2021.

[19] Mustin, 2021.

[20] U.S. Navy, Naval Sea Systems Command, "Aircraft Carriers—CVN," webpage, September 17, 2020.

because of their forward-deployed posture and ability to project force with their embarked aircraft. According to Naval Sea Systems Command, "the Nimitz and Gerald R. Ford-class aircraft carriers are the largest warships in the world, each designed for an approximately 50-year service life with just a single mid-life refueling."[21] The USS *Gerald R. Ford* (CVN 78) is the namesake of and first of the next generation of carriers.[22] (Note that, as of this writing, the USS *Gerald R. Ford* had not yet entered the fleet and was in the final stages of testing.) The carrier's real combat capability, though, resides in the carrier air wing, which accounts for about 1,500 personnel. However, what makes the wing truly unique from ground-based counterparts in other services is the platform that gives it its range. The aircraft carrier itself is employed with a crew of 3,000–3,200 sailors assigned to the vessel, who are known as "ship's company." [23]

As the largest warships in the world, aircraft carriers require a tremendous amount of maintenance. A substantial portion of this maintenance must occur in dry dock, at a shipyard, and often takes months. However, extremely high demand for aircraft carriers by GCCs, with approval by the Joint Staff and Secretary of Defense, has led to maintenance deferments, which have led to a maintenance debt over time. The debt requires other aircraft carriers to fill in, perhaps earlier than scheduled, thereby creating a vicious cycle. The Navy's response to resolve the ever-deepening maintenance debt was to create OFRP, discussed previously, to manage its force generation cycles so that it could address critical maintenance and manage fleet readiness.

As noted previously, the Navy addresses D2D and M2D differently from the other services, but the clear defining feature is its platform-centric construct. Other services rely on their vehicles and equipment to fulfill their assigned tasks, but none possess assets that are anything like capital ships with crews numbering in the thousands. As described earlier, from the Navy's perspective, D2D is an output of OFRP.[24]

Modeling Aircraft Carrier Availability

We analyzed historical carrier deployments in the context of a (notional) OFRP cycle. As noted earlier, OFRP is designed around a 36-month cycle, as shown in Figure 5.1. Each cycle begins with a maintenance phase followed by a training phase and ends with a sustainment phase, which includes one or more deployments.[25]

---

[21] U.S. Navy, Naval Sea Systems Command, 2020.

[22] U.S. Navy, Naval Sea Systems Command, 2020.

[23] U.S. Navy, Naval Sea Systems Command, 2020.

[24] This theme came up in multiple conversations with Navy officials.

[25] A standard OFRP cycle graphic would show four explicit phases—maintenance, training, deployment, and sustainment. However, because more than one deployment may occur during the latter two phases, we have chosen to graphically represent them as a single sustainment phase, during which deployments occur.

The length of each phase and the number of possible deployments during sustainment depends on the maintenance phase, as shown in Figure 5.2. Two types of maintenance can occur during an OFRP cycle—a pier-side planned incremental availability (PIA), which is notionally expected to last approximately seven months, or a dry-dock PIA (DPIA), which notionally lasts approximately 16 months.[26] During a PIA cycle, approximately eight months of training are required following the maintenance phase to achieve crew certification or recertification. Then, there is a 21-month sustainment phase during which carriers typically deploy once for seven months with the possibility of a second deployment.[27] The longer DPIA occurs every third cycle and necessitates a longer training period (approximately ten months) to achieve recertification because of the crew's extended time away from the ship. As a result, the sustainment period lasts only ten months, during which the carrier likely completes just a single (and slightly longer) deployment.

Using these theoretical PIA and DPIA cycles, we estimated D2D ratios based on a corresponding OFRP schedule. Assuming that the OFRP schedule began with a PIA cycle and that exactly one deployment per sustainment phase occurred at the beginning of the sustainment phase, we created a theoretical schedule, as shown in Figure 5.3. The figure shows approxi-

FIGURE 5.1
Optimized Fleet Response Plan

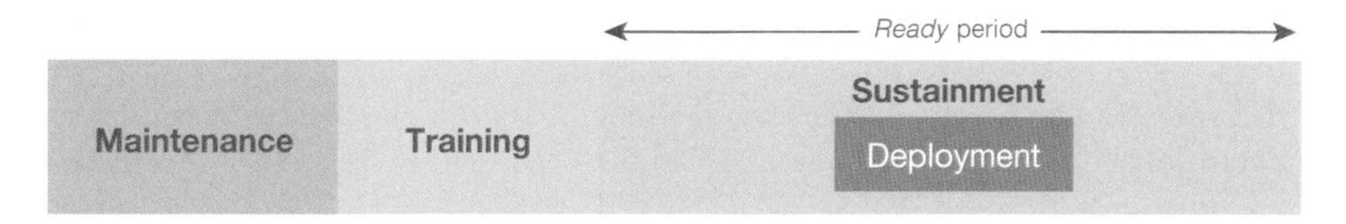

FIGURE 5.2
Optimized Fleet Response Plan with Phase Details

Months	1	2	3	4	5	6	7	8	9	10	11	12	13	14	15	16	17	18	19	20	21	22	23	24	25	26	27	28	29	30	31	32	33	34	35	36
PIA cycle	PIA							Training								Ready period (1–2 deployments)																				
DPIA cycle	DPIA																Training										Ready period (1 deployment)									

[26] Refueling and complex overhaul (RCOH) is a third type of maintenance, which occurs only once, at a carrier's midlife, and lasts nearly four years. This type of maintenance is not explicitly considered in our analysis.

[27] Should a carrier deploy twice during a given sustainment phase, it is not clear how the time between deployments would be considered. If it is considered dwell, then a severe violation of the D2D redline would occur during this single sustainment phase.

mately three and a half cycles to capture three D2D periods.[28] The middle (colored) row reflects the individual phases of the OFRP schedule, and the bottom row indicates the time frames for the corresponding deploy and dwell periods. For example, the deployment phase of the first PIA cycle represents the start of the first D2D period and lasts seven months. Following this deployment are 29 months of dwell, which correspond to the sustainment phase of the first PIA cycle and the maintenance and training phases of the second PIA cycle. During this period, the D2D ratio is approximately 1:4.1. We similarly computed the D2D ratios for the second and third D2D periods as 1:5.7 and 1:1.8, respectively. Across three cycles, the *average* D2D ratio is approximately 1:3.7, which is within the goal.[29]

This analysis suggests that carriers would, on *average*, meet the goal D2D ratios assuming ideal circumstances—that is, assuming that the OFRP cycles were adhered to exactly. However, as we alluded to previously, adherence to OFRP cycles has not always occurred. As a result of decades of overuse, aircraft carriers have been plagued with maintenance delays, and these delays have had widespread effects on the entire carrier fleet, affecting deployments across the board, as well as the fleet's ability to adhere to OFRP. Thus, we also examined historical deployment data to compare observed D2D ratios with the theoretical ones that we computed.

Figure 5.4 depicts carrier deployments from 2015 (the period during which initial implementation of OFRP was expected to begin) through June 2021. The USS *Ronald Reagan* and the USS *George Washington* are not reflected in the figure. The USS *Ronald Reagan* is part of the forward-deployed naval forces stationed at Yokosuka, Japan, and does not follow OFRP. Although the USS *George Washington* completed a deployment between September and December 2015, it did not deploy again before entering RCOH in August 2017, so there was no corresponding dwell period following the 2015 deployment. At the time of this analysis, the USS *George Washington* was still in RCOH.[30]

---

[28] Note that the PIA-PIA-DPIA cycle pattern will repeat beginning in year 10, so it is sufficient to examine the three and a half cycles shown here.

[29] We recognize that modification to the timing of any one deployment affects the D2D ratio of two cycles and that there are numerous possibilities for the timing of each of the four relevant deployments. To estimate a range for the possible D2D ratios, we examined the 16 variations of the theoretical OFRP cycle that results from each of the four deployment periods occurring at either the beginning or the end of the carrier's corresponding sustainment period. At the individual-cycle level, a minimum D2D ratio of 1.7 is observed when the DPIA cycle deployment occurs at the end of its sustainment period and the next deployment occurs at the beginning of its sustainment period. Similarly, a maximum D2D ratio of 6.1 is observed when the first PIA cycle deployment occurs at the beginning of its sustainment period and the next deployment occurs at the end of its sustainment period. Across OFRP, average D2D ratios ranging from 1:3.1 to 1:4.3 are observed under these deployment time assumptions, all within the goal. It might be possible to devise a deployment timeline that places deployments in the middle of their corresponding sustainment periods that would further increase the D2D ratio, possibly to the point of violating the goal ratio policy.

[30] Because D2D ratios are measured at the unit level, we did not seek or include data on crew turnover or continuity.

FIGURE 5.3

Theoretical Optimized Fleet Response Plan and Deployment-to-Dwell Ratios

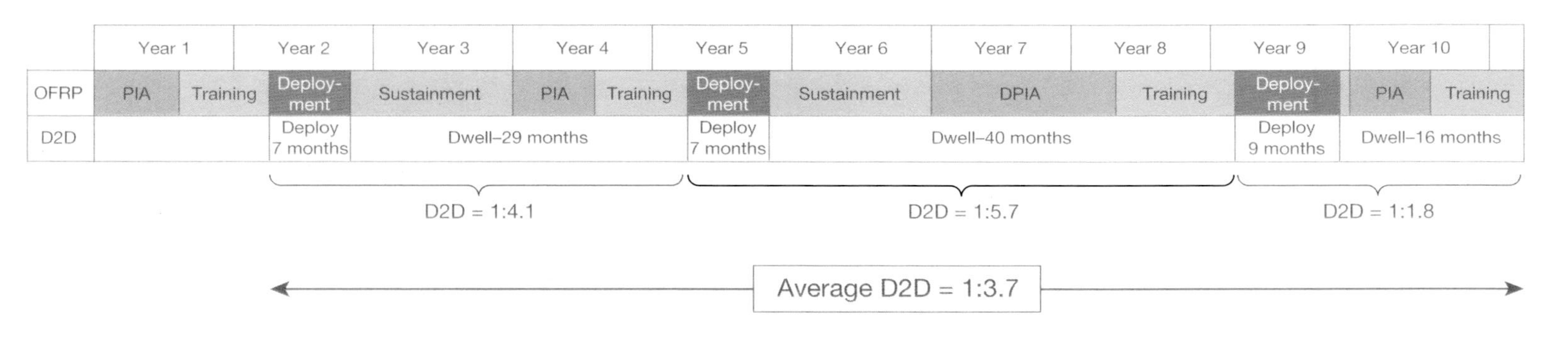

FIGURE 5.4

Navy Aircraft Carrier Deployments, 2015–2021

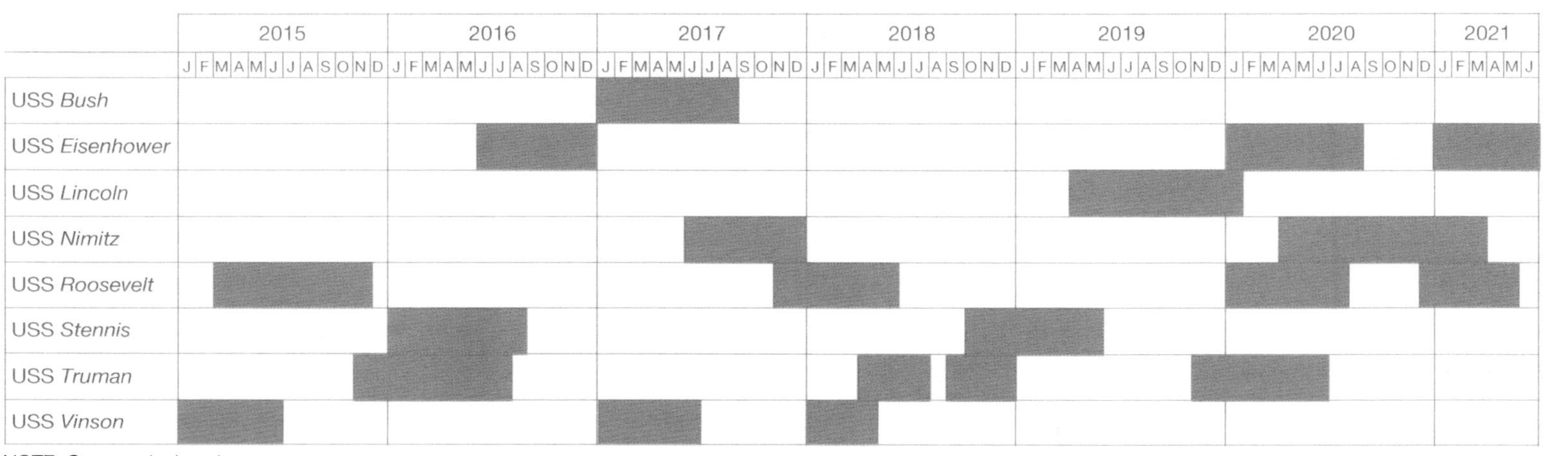

NOTE: Green = deployed.

Using the deployment chart in Figure 5.4, we identified the wide range of deployment and dwell times experienced by carriers and shown in Table 5.1. For example, between 2015 and 2021, carriers experienced deployments ranging from 3.4 months to 9.8 months, with an average (and median) of 6.8 months. More-significant differences were observed in dwell periods, which ranged from 1.3 months to 37.1 months. Short dwell periods fell significantly below the theoretical dwell periods of 16–40 months. Minimum and maximum deployed periods did not necessarily correspond to minimum and maximum dwell periods.

We also examined the D2D ratios for carriers. Across the 12 D2D periods observable in Figure 5.4, D2D ratios ranged from 0.4 (experienced by the USS *Truman* in 2018) to 5.25 (experienced by the USS *Eisenhower* between 2016 and 2020), with four such periods falling below the 1:2 threshold ratio and seven falling below the 1:3 goal ratio. On average, a D2D ratio of 2.6 was observed, which suggests that, *on average and over time*, carriers do meet the goal D2D ratio of 1:3. The D2D ratios for each of the eight carriers shown in Table 5.2 indicate

TABLE 5.1

Historic Deployment and Dwell for Navy Aircraft Carriers, 2015–2021

	Minimum	Maximum	Average	Median
Deploy (months)	3.4	9.8	6.8	6.8
Dwell (months)	1.3	37.1	17.3	20.0
D2D ratio[a]	0.4	5.3	2.6	2.7

[a] The D2D ratio statistics are based on observed D2D periods and do not necessarily align to the corresponding deploy and dwell times of the same statistics. For example, the maximum observed D2D ratio of 5.25 results from deployments of the USS *Eisenhower* in 2016 and 2020—a 2016 deployment period of 7.1 months followed by a 37.1-month dwell period. However, the maximum observed deployment period of 9.8 months was the 2019–2020 deployment of the USS *Lincoln*.

TABLE 5.2

Deployment-to-Dwell Ratios for Individual Aircraft Carriers, 2015–2021

Carrier	Minimum D2D	Maximum D2D	Average D2D
USS *Bush*[a]		N/A	
USS *Eisenhower*	1:0.9	1:5.3	1:3.1
USS *Lincoln*[a]		N/A	
USS *Nimitz*[b]		1:4.5	
USS *Roosevelt*	1:0.9	1:3.4	1:2.4
USS *Stennis*[b]		1:3.7	
USS *Truman*	1:0.4	1:3.1	1:2.0
USS *Vinson*	1:1.2	1:2.0	1:1.6

NOTES: N/A = not applicable.

[a] Only one deployment occurred during the 2015–2021 time frame, so no D2D ratio could be computed.

[b] Only two deployments (i.e., one dwell period) occurred during the 2015–2021 time frame, so only a single D2D ratio could be computed.

that most carriers have experienced at least one instance of a limited dwell period following a deployment over the past six years.

Historic deployment data reveal that carriers are deploying more frequently than suggested by OFRP but are still nearly meeting goal ratios *on average and over time* when considered as a fleet—average and median D2D ratios are just under 1:3. As shown in Table 5.2, D2D periods are quite different for individual ships. Twenty-five percent of the ships considered had deployed only once during the six-year period, so corresponding D2D ratios could not be calculated. Another 25 percent experienced only two deployments, so only a single D2D period was available for consideration, although both carriers met the goal D2D ratio. The remaining 50 percent experienced at least two deployments, and three of the four ships saw both extremely low D2D ratios (less than 1:1) and sufficiently high D2D ratios (greater than 1:3).

The long-term impact of OFRP on D2D is still unknown because carriers have not been operating under OFRP long enough to collect sufficient data. Carriers were originally slated to begin implementation of OFRP in FY 2015, but a significant maintenance delay experienced by the USS *Eisenhower* in 2015 affected the implementation across the fleet (see Figure 5.5). What was expected to be a 15-month maintenance period turned into 32 months, tying up shipyard resources and causing additional deferments and delays in carrier maintenance.

FIGURE 5.5
Carrier Implementation of Optimized Fleet Response Plan

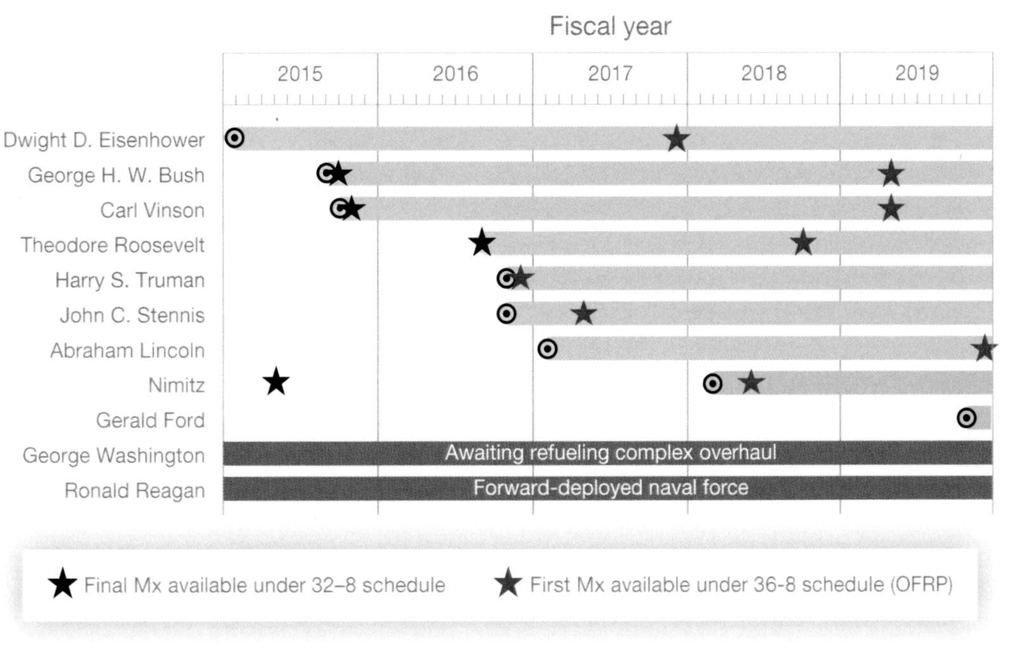

SOURCE: Adapted from U.S. Government Accountability Office, *Navy's Optimized Fleet Response Plan: Information Provided to Congressional Committees*, Washington, D.C., GAO-16-466R, 2016.
NOTE: Mx = mission.

The USS *Truman* was the first carrier to begin a maintenance availability under OFRP, in September 2016, and the USS *Lincoln* was the last carrier to enter OFRP, in September 2020. At the time of this analysis, no carrier had completed a second maintenance availability under OFRP. Persistent maintenance deferments and delays continue to challenge the Navy's ability to adhere to the OFRP schedule. In addition to the current limited operational time under OFRP, available data are insufficient to assess the impact of OFRP on D2D.

Summary

As the modeling results demonstrated, historical data suggest that carriers are deploying more frequently than OFRP was potentially designed to allow, and, although current deployments appear to meet goal ratios *on average and over the period of this analysis*, the minimum D2D ratio identified is well below the threshold ratio required by policy. The limited operational time under OFRP and the persistent maintenance deferments challenge both the Navy's ability to adhere to the schedule and our ability to assess how D2D is affected by operational and maintenance demands and by PERSTEMPO limitations.

However, ongoing demand for carrier deployments, modernization requirements, and persistent maintenance delays raise questions about D2D in the long term. There is ongoing debate about whether combatant command demand for carriers is supportable. Unprogrammed sea time compounds existing and very real issues with required maintenance and attempts to modernize the fleet, while shipyard delays due to various factors have dramatically and negatively affected ship availability in recent years. As a 2015 Center for Strategic and Budgetary Assessments report noted,

> The last two decades have been busy ones for the Navy. Between 1998 and 2014, the number of ships deployed overseas remained roughly constant at 100. The fleet, however, shrank by about 20 percent. As a result, each ship is working harder to maintain the same level of presence.[31]

In addition, as an April 2021 congressional letter to Secretary of Defense Lloyd Austin noted,

> The 'tyranny of the now' is wearing out man and machine at too high a rate to ensure success both now and later. Future readiness can no longer be sacrificed at the altar of lower-priority requirements. All the RFFs [requests for forces], and their approvals, represent more than just a failure to adhere to the existing GFMAP; they also reflect a failure to prioritize. The consistent high operational tempo of our military assets is creating unsustainable deploy-to-dwell ratios. Put plainly, regular circumvention of the GFMAP is leaving the services scrambling at a time when they need to rebuild the health of the force.

[31] Bryan Clark and Jesse Sloman, *Deploying Beyond Their Means: America's Navy and Marine Corps at a Tipping Point*, Washington, D.C.: Center for Strategic and Budgetary Assessments, 2015, p. 6.

> At this rate, the desire to solve every immediate problem, regardless of its strategic prioritization, may hollow the force for the next generation. It is imperative that the COCOMs [combatant commanders] accept and share the appropriate amount of risk required to balance their needs against the chiefs' requirement to recruit, train and modernize the services in the long term.[32]

The Navy's ability to meet D2D/M2D ratios *on average and over time* might not last, and it might depend on the period analyzed and the unique factors at work during that period. Given the modeling observations described in this chapter, along with demand for the active and reserve components, consideration of how needed maintenance for the Navy's carrier fleet, modernization demands, and increasing operational requirements might affect D2D/M2D outcomes for the Navy will be critical as time progresses. The limited data under OFRP point to areas of concern, but additional study is needed.

---

[32] Robert J. Wittman, Seth Moulton, Michael R. Turner, Jackie Speier, Doug Lamborn, Elise Stefanik, Joe Wilson, Don Bacon, Jack Bergman, Mo Brooks, Kaiali'i Kahele, Van Taylor, Scott Desjarlais, and Blake Moore, letter to Lloyd J. Austin, Secretary of Defense, and Kathleen Hicks, Deputy Secretary of Defense, Washington, D.C., April 5, 2021, p. 2.

CHAPTER SIX

Dimensions of Policy Implementation and Recommendations

The analysis of the JFEs selected for this study, along with the examination of how the services approach D2D/M2D policy, revealed insights that help characterize how this policy manifests across DoD. In this chapter, we present these insights according to six dimensions that illustrate policy implementation across the services:

- alignment of service policy with OSD policy
- service management of D2D/M2D policy
- services' ability to assess the availability of JFEs to meet emerging requirements and the impacts on readiness
- JFE compliance with D2D/M2D goal and threshold ratios
- policy implications for reserve component use
- policy influence on force structure decisions.

The chapter concludes with recommendations for the future of D2D/M2D policy.

Dimensions of Deployment-to-Dwell/Mobilization-to-Dwell Policy Implementation

Alignment of Service Policy with Office of the Secretary of Defense Policy

Service policies generally align with the OSD policy, with additional service-specific elements in some cases, as shown in Table 6.1. The Army relies solely on the OSD policy and has no service-specific policy. The USAF relies largely on the OSD policy but augments it with internal goals for D2D and M2D by weapon system or support area. The Marine Corps has a service-specific policy with added elements requiring assessment of D2D/M2D ratios for individual marines and specified processes for unit compliance or waivers of D2D/M2D ratios. Marine Corps policy also excludes exercises from being counted as operational deployments. Finally, the Navy's policy aligns with the OSD policy, with two additional elements. Deployment length and cumulative underway days are tracked, and there is one exception to

TABLE 6.1
Service Deployment-to-Dwell/Mobilization-to-Dwell Policy Alignment with Office of the Secretary of Defense Policy

Service	Observations
Army	Relies on OSD policy, no separate service-specific policy
Air Force	Relies on OSD policy, service-specific policy covers mobilization • In practice, sets internal goals by weapon system or support area
Marine Corps	OSD policy is incorporated in service policy, with additional elements • Requires assessment of individual marines' D2D/M2D ratios • Describes processes for unit compliance or waivers • Specifically excludes exercises from operational deployments
Navy	Service policy aligns with OSD policy; provides additional elements • Deployment length and cumulative underway days are tracked • Multicrewed forces designed for continuous presence are excepted

D2D/M2D policy for multicrewed forces designed for continuous presence. In this case, the threshold ratio for these crews is 1:1.

Service Management of Deployment-to-Dwell/Mobilization-to-Dwell Policy

Service management of D2D/M2D policy varies considerably across the services. This variation, as detailed in Table 6.2, is largely attributed to the principal drivers of D2D/M2D management within the service. Specifically, management primarily focuses on one of three things: the unit, the platform, or the individual. The Army largely manages D2D and M2D at the unit level. Force managers maintain D2D/M2D for unit types according to their level of employment with the GFMAP or emerging requirement. This methodology makes sense for the Army because most of its requirements are for a specified unit. Unit personnel officers and branch managers optimize for individual soldier circumstances when a soldier's D2D/M2D ratio is at risk of breaking the threshold in the wake of broader, unit-level force management decisions.

The USAF's management of D2D and M2D—in stark contrast to the Army's—involves tracking the deployment history of individuals rather than units to decide who to task for missions. Consequently, PERSTEMPO also plays a significant factor in individual sourcing decisions for USAF deployments. In cases of unit deployments, force managers in the USAF source units with the individuals who are the most eligible (according to deployment history and PERSTEMPO) from the larger garrison unit. Additionally, the MAF are not strictly implementing the D2D/M2D policy as written. They use T2D as the primary metric for the daily management of OPTEMPO, which could become an alternative metric for capturing deployment and dwell times to reflect the stress of this force element.

Marine Corps policy directs yet another approach in which D2D and M2D are managed by continuously tracking unit and individual D2D/M2D ratios. Finally, the Navy has

TABLE 6.2
Service Management of Deployment-to-Dwell/Mobilization-to-Dwell Policy

Service	Observations
Army	• The **unit** is the primary means for measuring D2D/M2D. • Units and personnel managers optimize for individual soldier circumstances.
Air Force	• The USAF tracks individual deployment history; PERSTEMPO is a factor in selecting individuals for deployment. • Unit deployments are sourced from the most eligible individuals from garrison units.
Marine Corps	• **Unit** and **individual** D2D/M2D ratios are continuously tracked.
Navy	• **Platform-** and **unit-centric focus for the active component:** OFRP maintenance schedules drive force generation and ensure sufficient dwell. • **M2D is managed by the individual**; most mobilizations are IAs. • USNR unit mobilizations are limited to a few unit types, but unit members are mobilized as individuals.

a composite approach to D2D/M2D management according to component. The active component is managed from a platform- and unit-centric focus. Although D2D/M2D ratios for individual sailors are maintained, the OFRP maintenance schedules, which ensure sufficient dwell for sailors, are the primary driver of naval force generation. For the reserve component, the Navy manages M2D for individual sailors, who are mobilized almost exclusively as IAs. USNR unit mobilizations have been restricted to a few unit types, including Navy Special Warfare units, Strike Fighter squadrons, Electronic Attack squadrons, Fleet Logistics Support squadrons, Patrol squadrons, Fighter Composite units, Navy Mobile Construction battalions, Expeditionary Logistics, and Coastal Riverine squadrons. In cases of unit mobilizations, the Navy typically mobilizes unit members as individuals.

Services' Ability to Assess the Availability of Joint Force Elements to Meet Emerging Requirements and the Impacts on Readiness

Interviews with individuals in DoD's force management community focused in part on understanding service models and methods for assessing impacts to operational readiness (the readiness required to respond to a major contingency) considering D2D/M2D policy. Generally, the services tend to focus on the near term (12–18 months) when assessing the impacts of fulfilling an emerging requirement on operational readiness. We found no unified modeling approach within the services to conduct such assessments beyond an aggregated long-term projection based on D2D/M2D ratios. Notably, we found that the Marine Corps' active component infantry battalions are scheduled to be employed at the threshold D2D/M2D ratio, which implies reliance on the reserve component for fulfilling an enduring requirement or repurposing an active component unit's deployment. Service-specific observations with respect to modeling JFE availability to meet an emerging requirement and the associated impacts to operational readiness are shown in Table 6.3.

TABLE 6.3

Service Modeling of Joint Force Element Availability and Impacts on Readiness

Service	Observations
Army	No existing model; JFE availability is relatively transparent and easy to assess in the short term
Air Force	There does not appear to be a unified model that examines impacts to readiness for meeting deployment requirements
Marine Corps	No existing model; JFE studied in this research is scheduled to be employed at threshold D2D/M2D level
Navy	Routinely examines prospective impacts to operational schedules; D2D/M2D impacts are considered but are not a primary input for assessments

Joint Force Element Compliance with Deployment-to-Dwell/ Mobilization-to-Dwell Goal and Threshold Ratios

The modeling and analysis that we conducted indicate that compliance with D2D/M2D goal and threshold ratios varied across the services for the four JFE case studies, as shown in Table 6.4. Our analysis shows that all JFEs were capable of meeting demand under the scenarios examined while being compliant with the threshold D2D/M2D ratios. In contrast, JFEs were largely unable to meet demands in these scenarios while following the goal D2D/ M2D ratios. The Army's ABCT, USAF's KC-135s, and the Marine Corps' infantry battalion could not meet requirements while following the goal ratios, resulting in several consecutive months of missed operational requirements. Finally, although the Navy's carriers complied with goal and threshold ratios, as shown in Table 6.4, this assessment is based on average D2D ratios for the time frame examined. As mentioned in our assessment of Navy carriers, in some instances carriers port for a short period (one to two months) before departing again, which results in breaking the threshold ratio unless an average ratio is calculated over a longer period. Broadly speaking, the services' inability to meet requirements at goal D2D/

TABLE 6.4

Joint Force Element Compliance with Deployment-to-Dwell/Mobilization-to-Dwell Goal and Threshold Ratios

	Goal D2D/M2D Ratio		Threshold D2D/M2D Ratio	
JFE—Service	Active	Reserve	Active	Reserve
ABCT—Army				
KC-135—Air Force				
IN Bn—Marine Corps				
Carrier—Navy		N/A		N/A

NOTES: IN Bn = infantry battalion; N/A = not applicable. Green = demand met; yellow = demand partially met; red = demand not met. The Army and Marine Corps JFEs failed to meet demand under goal ratios. The Air Force KC-135 assessment is based on percentage of demand met; 82 percent of demand was met under the threshold D2D/M2D ratios, and 67 percent of demand was met under the goal D2D/M2D ratios. Navy carriers met D2D ratios on average during the time frame examined, influenced in part by maintenance delays.

M2D ratios may be driven by a lack of Secretary of Defense enforcement of compliance at the goal ratios.

Policy Implications for Reserve Component Use

Use of the reserve components is unique to each service, as the observations in Table 6.5 indicate. The selected JFE modeling and interviews with service force managers indicate that reserve component use varies from a focus on individuals and how requirements for those individuals are met to a focus on high-interest force elements like the ABCT. Accordingly, policy implications for reserve component use are quite broad.

In the Army, use of the reserve component is critical in fulfilling known and emerging requirements. Thus, Army reserve units fulfill the role of operational reserve for a portion of the Army's formations. Performing this role potentially limits the force structure available to meet surge demands for a major contingency operation. In contrast, USAF mobilizations are largely filled by volunteers, with little effect on the USAF's ability to fulfill requirements in a surge scenario. The Marine Corps future force structure plans for the active component as described in *Force Design 2030* would reduce the number of active infantry battalions. This change in force structure would either (1) create a risk in fulfilling current requirements while abiding by threshold ratios if requirements continue to be met by the active component or (2) potentially change the active and reserve component mix for this unit type. Navy use of the reserves is largely restricted to individuals, but Navy leadership indicates that nonmaritime demands for IA have adversely affected the Navy's ability to meet surge requirements for a major contingency.

TABLE 6.5

Reserve Component Use and Deployment-to-Dwell/Mobilization-to-Dwell Policy Implications

Service	Observations	Policy Implications
Army	Reserve component units are critical in fulfilling requirements	• Affects strategic reserve surge capacity
Air Force	A significant portion of reserve component participation is voluntary	• Has little effect on ability to fulfill requirements
Marine Corps	*Force Design 2030* calls for goal ratios and reductions in unit types	• Creates risk in fulfilling current requirements • Potentially changes active and reserve component mix
Navy	Reserve component use is largely restricted to individuals	• Nonmaritime IA demands have adversely affected strategic reserve

NOTE: We define *strategic reserve* as those units or individuals required to respond to a major contingency operation.

Policy Influence on Force Structure Decisions

We found that D2D/M2D policy has an impact on both Army and Marine Corps force structure decisions but does not appear to influence these decisions for the USAF and the Navy, as shown in Table 6.6. The Army's process for evaluating force mix from year to year incorporates D2D/M2D ratios. The set of parameters that the Army uses to describe unit availability incorporates D2D/M2D ratios among many other factors. Resulting force mix recommendations reflect D2D/M2D policy to some degree. Recent guidance from the Commandant of the Marine Corps to return the active component to a 1:3 ratio implies that fewer requirements can be met, use of the reserve component must be increased to meet current requirements, or the active component must be restructured to meet requirements.

Broader Findings Across the Department of Defense

Our interviews with force management communities across DoD and our JFE modeling revealed additional findings that do not fit into the previously discussed dimensions of D2D/M2D policy practice and impacts:

- *The services report that they rarely push back or reclama emerging requirements.* This might be an indication that they are incentivized to maximize unit, personnel, or platform employment—demonstrating service value. However, the services do not provide DoD an assessment of the long-term impacts to operational readiness when filling these requirements.
- *When units cross the D2D/M2D threshold, the mission requirement tends to receive a Secretary of Defense waiver.*
- *The services have different views on what constitutes a deployment under D2D/M2D policy.* The following are specific examples of this:
  - For some force elements, the USAF asserts that D2D and M2D do not capture activities similar to operational deployments; for other force elements typically deployed in place, operational deployments do not count for D2D.

TABLE 6.6

Deployment-to-Dwell/Mobilization-to-Dwell Policy Influence on Force Structure Decisions

Service	Observations
Army	Plays a role in determining unit availability during contingencies; informs overall unit mix
Air Force	Does not appear to influence force structure decisions
Marine Corps	Marine Corps Commandant guidance to return the active component to 1:3 D2D implies that fewer requirements can be met, use of the reserve component must be increased, or the active component must be restructured to meet requirements
Navy	Does not appear to influence force structure decisions

  - Marine Corps forward presence rotational deployments count as operational deployments under D2D/M2D policy.
  - It is unclear how the Navy counts ships returning to port for one to three months between two deployment periods. Hence, for certain periods, these platforms may break goal or threshold ratios because of the quick turnaround.
- *D2D/M2D processes are not directly related to readiness.* We observed that current threshold and objective D2D/M2D ratios are independent of Army readiness objectives while the Navy's OFRP cycle was more indicative of readiness than the D2D metric.
- *The policy appears to do no harm but does not fully address or accomplish its* ***stated*** *objectives across the force.* During our research, several stakeholders emphasized the fact that a purpose of the newly drafted D2D/M2D policy is to ensure appropriate operational readiness in light of DoD's renewed focus on a return to great-power competition. Force implementation involves a much broader array of considerations for D2D/M2D management than the unit and the individual; platforms and composite units that are needed for response to major contingencies are also important drivers for force management. Hence, the policy should specify platforms and composite units if the stated objective of ensuring appropriate operational readiness is a primary goal of the policy.

Recommendations

Three recommendations emerged from our investigations into the services' current approaches, policies, and models for D2D/M2D management and the JFE case studies: Clarify the purpose of the D2D/M2D policy, consider redefining what counts as a deployment, and review how D2D and M2D are managed.

Clarify the Purpose of the Policy Within the Current Operational Context

The operational context has fundamentally changed since the original D2D/M2D policy was developed. Although the policy was originally intended to prevent the overexposure of individuals to combat (which is still written in the policy), services track all operational deployments as part of their D2D/M2D calculations rather than focusing on only those that result in combat exposure. Additionally, throughout the conduct of this study, a stated purpose of D2D/M2D policy was to preserve unit readiness. Current D2D/M2D policy assumes that all deployments consume readiness; in practice, some deployments may contribute to readiness. The current operational context does not include large numbers of the force being exposed to combat conditions as frequently as when the policy was developed and offers an opportunity to revisit the purpose of the policy so that it fully suits DoD needs.

Consider Redefining What Counts as a Deployment

The current policy does not capture activities that directly support (but do not officially fulfill) an approved Secretary of Defense operational requirement that requires service members to be away from home, which might lead to false assumptions on available capacity for responding to a major contingency operation. Additionally, we found that some units' D2D/M2D ratios might be artificially inflated. Conditions for this exist when a Secretary of Defense–approved requirement coincides with a routinely performed unit rotational commitment. With a return to great-power competition and renewed focus on readiness required to respond to a major contingency, D2D/M2D policy could allow services to make judgments about excluding readiness-enhancing deployments from D2D/M2D calculations. At the time of writing, DoD is considering changes in what counts as a deployment.

Review How Deployment to Dwell and Mobilization to Dwell Are Managed Across the Department of Defense

This research revealed that the services have different ways of managing the force that are based in part on D2D/M2D policy. The Army's management of D2D and M2D focuses solely on the unit. In some cases, the nature of requirements for the USAF necessitates careful management of the individual, the crew, and the platform, while the Navy largely manages the platform and the individual. D2D/M2D tracking and corresponding reporting requirements should consider platform, crew, and unit composition for fully understanding the effects of deployments on operational readiness.

Finally, if the purpose of D2D/M2D policy remains as written—to avoid the overexposure of individuals to operational deployment—then the recently revised DoD PERSTEMPO policy could suffice in fulfilling this purpose. However, if the purpose of the policy is to preserve operational readiness, DoD should consider D2D/M2D tracking only for those JFEs and other essential force elements needed for response to major contingency operations. Such a move would dramatically decrease the complexity of D2D/M2D tracking and provide greater clarity into the effects that emerging requirements have on DoD's ability to respond to major contingencies.

Abbreviations

ABCT	armored brigade combat team
AMC	Air Mobility Command
AST	Army Synchronization Tool
BOG	boots on ground
CVN	aircraft carrier
D2D	deployment to dwell
DoD	Department of Defense
DPIA	dry-dock planned incremental availability
FORSCOM	U.S. Army Forces Command
GCC	geographic combatant command
GFMAP	Global Force Management Allocation Plan
IA	individual augmentee
JFE	joint force element
M2D	mobilization to dwell
MAF	Mobility Air Forces
MAFORGEN	Mobility Air Forces Force Generation
MARADMIN	Marine Corps Administrative Message
MARFORCOM	Marine Forces Command
MEU	Marine expeditionary unit
MPAFF	Multi-Period Assessment of Force Flow
OFRP	Optimized Fleet Response Plan
OPNAV	Office of the Chief of Naval Operations
OPNAVINST	Office of the Chief of Naval Operations Instruction
OPTEMPO	operational tempo
OSD	Office of the Secretary of Defense
PAA	primary aircraft authorized
PERSTEMPO	personnel tempo
PIA	planned incremental availability
RCOH	refueling and complex overhaul
T2D	task to dwell
TAA	Total Army Analysis
URC	unit readiness cycle
USAF	U.S. Air Force
USD P&R	Under Secretary of Defense for Personnel and Readiness

USNR	U.S. Navy Reserve
USTRANSCOM	U.S. Transportation Command

References

AR—*See* Army Regulation.

Army Regulation 220–1, *Army Unit Status Reporting and Force Registration—Consolidated Policies*, Washington, D.C.: Headquarters, Department of the Army, April 15, 2010.

Army Regulation 525–29, *Force Generation—Sustainable Readiness*, Washington D.C.: Headquarters, Department of the Army, October 1, 2019.

Best, Katharina Ley, Igor Mikolic-Torreira, Rebecca Balebako, Michael Johnson, Trung Tran, and Krista Romita Grocholski, *Assessing Force Sufficiency and Risk Using RAND's Multi-Period Assessment of Force Flow (MPAFF) Tool*, Santa Monica, Calif.: RAND Corporation, RR-1954-A, 2019. As of August 18, 2021:
https://www.rand.org/pubs/research_reports/RR1954.html

Bonds, Timothy M., Dave Baiocchi, and Laurie L. McDonald, *Army Deployments to OIF and OEF*, Santa Monica, Calif.: RAND Corporation, DB-587-A, 2010. As of August 21, 2021:
https://www.rand.org/pubs/documented_briefings/DB587.html

Castro, Carl A., and Amy B. Adler, "OPTEMPO: Effects on Soldier and Unit Readiness," *Parameters*, Vol. 29, No. 3, Autumn 1999, pp. 86–95.

———, *The Impact of Operations Tempo: Issues in Measurement*, Ft. Detrick, Md.: U.S. Army Medical Research and Material Command, 2000.

Chu, David, Under Secretary of Defense for Personnel and Readiness, "Measuring Boots on Ground (BOG)—Snowflake," memorandum for the Secretary of Defense, Washington, D.C., November 22, 2004.

Clark, Bryan, and Jesse Sloman, *Deploying Beyond Their Means: America's Navy and Marine Corps at a Tipping Point*, Washington, D.C.: Center for Strategic and Budgetary Assessments, 2015.

Deni, John R., *Rotational Deployments vs. Forward Stationing: How Can the Army Achieve Assurance and Deterrence Efficiently and Effectively?* Carlisle Barracks, Pa.: U.S. Army War College Press, August 25, 2017.

Department of the Air Force, "AFFORGEN Update: Sustainable Force Offerings," briefing, January 25, 2021.

Department of the Army Pamphlet 220–1, *Defense Readiness Reporting System—Army Procedures*, Washington, D.C., Headquarters, Department of the Army, November 16, 2011.

DoD—*See* U.S. Department of Defense.

Everstine, Brian W., "CSAF Plans a Better Deployment Model," *Air Force Magazine*, October 1, 2020.

Gates, Robert M., Secretary of Defense, "Utilization of the Total Force," memorandum for the Secretaries of the Military Departments, Chairman of the Joint Chiefs of Staff, and Under Secretaries of Defense, January 19, 2007.

Headquarters, Department of the Army, *Regionally Aligned Readiness and Modernization Model (ReARMM)*, Washington, D.C., forthcoming.

Hosek, James, *Perstempo: Does It Help or Hinder Reenlistment?* Santa Monica, Calif.: RAND Corporation, RB-7532, 2004. As of August 22, 2021:
https://www.rand.org/pubs/research_briefs/RB7532.html

Kapp, Lawrence, *Recruiting and Retention in the Active Component Military: Are There Problems?* Washington, D.C., Congressional Research Service, RL31297, February 25, 2002.

MARADMIN—*See* Marine Corps Administrative Message.

Marine Corps Administrative Message 346/14, "Deployment-to-Dwell, Mobilization-to-Dwell Policy Revision," Washington, D.C.: U.S. Marine Corps, July 14, 2014.

Marine Corps Reference Publication 1-10.1, *Organization of the United States Marine Corps*, Washington, D.C.: U.S. Marine Corps, August 26, 2015.

MCRP—*See* Marine Corps Reference Publication.

Modly, Thomas B., Acting Secretary of the Navy; Michael M. Gilday, Chief of Naval Operations; and David H. Berger, Commandant of the U.S. Marine Corps, "Fiscal Year 2021 Department of the Navy Budget," statement before the Senate Armed Services Committee, U.S. Senate, Washington, D.C., March 5, 2020.

Mustin, John B., Chief of the U.S. Navy Reserve, "Fiscal Year 2022 National Guard and Reserve," statement before the Subcommittee on Defense, Committee on Appropriations, U.S. Senate, Washington, D.C.: May 18, 2021.

Office of the Chief of Naval Operations Instruction 3000.13E, *Navy Personnel Tempo and Operating Tempo Program*, Washington, D.C.: Department of the Navy, January 27, 2021.

Office of the Chief of Naval Operations Instruction 3000.15A, *Optimized Fleet Response Plan*, Washington, D.C.: Department of the Navy, November 10, 2014.

Office of the Under Secretary of Defense, "Under Secretary of Defense (Personnel & Readiness) Deployment-to-Dwell, Mobilization-to-Dwell Policy Revision," memorandum for Secretaries of the Military Departments and Chairman of the Joint Chiefs of Staff, Washington, D.C., November 1, 2013.

"Operations in South China Sea Increase Readiness of Squadrons: US Navy," *Business Standard*, last updated July 6, 2020.

OPNAVINST—*See* Office of the Chief of Naval Operations Instruction.

Paxton, John, Assistant Commandant of the Marine Corps, "U.S. Marine Corps Readiness," statement before the Subcommittee on Readiness, Committee on Armed Services, U.S. Senate, Washington, D.C., March 15, 2016.

Public Law 112–81, National Defense Authorization Act for Fiscal Year 2012, December 31, 2011.

U.S. Code, Title 10, Armed Forces; Subtitle A, General Military Law; Part II, Personnel, Chapter 50, Miscellaneous Command Responsibilities; Section 991, Management of Deployments of Members and Measurement and Data Collection of Unit Operating and Personnel Tempo, January 1, 2021.

U.S. Code, Title 37, Pay and Allowances of the Uniformed Services; Chapter 7, Allowances; Section 436, High-Deployment Allowance: Lengthy or Numerous Deployments; Frequent Mobilizations, 1999.

U.S. Department of Defense, *Quadrennial Defense Review Report*, Washington, D.C., September 30, 2001.

U.S. Government Accountability Office, *Navy's Optimized Fleet Response Plan: Information Provided to Congressional Committees*, Washington, D.C., GAO-16-466R, 2016.

U.S. Marine Corps, *Force Design 2030*, Washington, D.C., March 2020.

U.S. Marine Corps, *Force Design 2030: Annual Update*, Washington, D.C., April 2021.

U.S. Marine Corps Forces, Europe and Africa, "U.S. Marines and Sailors from Marine Rotational Force-Europe Complete Deployment to Northern Norway, Return to Camp Lejeune, N.C.," Defense Visual Information Distribution Service, April 16, 2021.

U.S. Navy, Naval Sea Systems Command, "Aircraft Carriers—CVN," webpage, September 17, 2020. As of July 1, 2021:
https://www.navy.mil/Resources/Fact-Files/Display-FactFiles/Article/2169795/aircraft-carriers-cvn/

USTRANSCOM—*See* U.S. Transportation Command.

U.S. Transportation Command, J3, "Policy Proposal for the Addition of Task-to-Dwell (T2D) as Tempo and Readiness Metric for USTRANSCOM-Assigned Mobility Air Forces (MAF)," memorandum to the Under Secretary of Defense for Personnel and Readiness, July 30, 2020.

Winkler, John D., "Developing an Operational Reserve: A Policy and Historical Context and the Way Forward," *Joint Forces Quarterly*, Vol. 59, 4th Quarter 2010, pp. 14–20.

Wittman, Robert J., Seth Moulton, Michael R. Turner, Jackie Speier, Doug Lamborn, Elise Stefanik, Joe Wilson, Don Bacon, Jack Bergman, Mo Brooks, Kaiali'i Kahele, Van Taylor, Scott Desjarlais, and Blake Moore, letter to Lloyd J. Austin, Secretary of Defense, and Kathleen Hicks, Deputy Secretary of Defense, Washington, D.C., April 5, 2021.

Wood, Dakota L., "An Assessment of U.S. Military Power: U.S. Marine Corps," Heritage Foundation, November 17, 2020.

Wright, Jessica L., Acting Under Secretary of Defense for Personnel and Readiness, "Under Secretary of Defense (Personnel & Readiness) Deployment-to-Dwell, Mobilization-to-Dwell Policy Revision," memorandum for Secretaries of the Military Departments and Chairman of the Joint Chiefs of Staff, Washington, D.C., November 1, 2013.